T0243079

Monograph Series   **Volume 1**

The Work of Mathematics Teacher Educators

Exchanging Ideas for Effective Practice

Edited by

Tad Watanabe
The Pennsylvania State University

Denisse R. Thompson
University of South Florida

Association of
Mathematics Teacher Educators

Published by the Association of Mathematics Teacher Educators

Headquarters:
San Diego State University
c/o Center for Research in Mathematics and Science Education
6475 Alvarado Road, Suite 206
San Diego, CA 92120

www.amte.net

Library of Congress Cataloging-in-Publication Data

The work of mathematics teacher educators : exchanging ideas for
effective practice / edited by Tad Watanabe, Denisse R. Thompson.
    p. cm. — (Monograph series ; v. 1)
  Includes bibliographical references.
  ISBN 978-1-62396-941-7
  1. Mathematics—Study and teaching—United States. 2. Mathematics
teachers—Training of—United States. I. Watanabe, Tad, 1959- II.
Thompson, Denisse Rubilee, 1954- III. Series: Monograph series
(Association of Mathematics Teacher Educators) ; v. 1.

  QA11.2.W67 2004
  510'.71—dc22

                            2004016674

The publications of the Association of Mathematics Teacher Educators
present a variety of viewpoints. The views expressed or implied in this
publication, unless otherwise noted, should not be interpreted as official
positions of the Association.

Contents

Preface   v

1. **Mathematics Teacher Education:**   1
    At a Crossroad

    Glenda Lappan, Michigan State University
    Kelly Rivette, Michigan State University

2. **Collaborative Efforts to Improve the**   19
    Mathematical Preparation of Middle Grades
    Mathematics Teachers: Connecting Middle
    School and College Mathematics

    James E. Tarr, University of Missouri-Columbia
    Ira J. Papick, University of Missouri-Columbia

3. **Conducting Research and Solving Problems:**   35
    The Russian Experience of Inservice Training

    Alexander Karp, Teachers College, Columbia University

4. **Using Alternative Assessment to Affect**   49
    Preservice Elementary Teachers' Beliefs
    About Mathematics

    David Charles Coffey, Grand Valley State University

5. **The Experiences in a Concepts in Calculus**   67
    Course for Middle School Mathematics Teachers

    Susann M. Mathews, Wright State University

6. **Drag, Drag, Drag: The Impact of Dragging**   87
    on the Formulation of Conjectures within
    Interactive Geometry Environments

    José N. Contreras, University of Southern Mississippi
    Armando M. Martinez-Cruz, California State University,
    Fullerton

7. **Spherical Geometry as a Professional Development**   103
    Context for K-12 Mathematics Teachers

    Janet Sharp, Montana State University

8. **Preparing for the Future: An Early Field Experience**   119
    that Focuses on Students' Thinking

    Laura R. Van Zoest, Western Michigan University

9. **Preparing Teachers to Teach Mathematics**   135
    Within a Constructivist Framework:
    The Importance of Listening to Children

    Beatriz S. D'Ambrosio, Indiana University
    Purdue University Indianapolis

10. **Revealing Current Practice Through Audio-Analysis**   151
    Releases the Power of Reflection to Improve Practice

    Ann R. Taylor, Southern Illinois University Edwardsville
    Barbara D. O'Donnell, Southern Illinois University
    Edwardsville

11. **The Interview Assignment: Evaluating a**   169
    Teacher Candidate's Knowledge of Mathematics
    Content, Questioning, and Assessment

    Patricia S. Moyer-Packenham, George Mason University

12. **Professional Development for Teacher and**   189
    Teacher Educator Through Sustained
    Collaboration

    Amy Roth McDuffie, Washington State University Tri-Cities
    Martha Mather, Washington State University Tri-Cities
    Karen Reynolds, Washington State University Tri-Cities

13. **Helping Teachers Develop New Conceptualizations**   205
    About the Teaching and Learning of Mathematics

    Roberta Y. Schorr, Rutgers, the State University
    of New Jersey Campus at Newark

Preface
AMTE Monograph 1
The Work of Mathematics Teacher Educators
©2004, pp. v-viii

Preface

This monograph marks a major step in the growth of the Association of Mathematics Teacher Educators (AMTE) as the leading organization for mathematics teacher educators. At the 2003 annual meeting in Atlanta, we celebrated the tenth anniversary of our organization, and the 2004 meeting in San Diego was the eighth AMTE annual meeting. This monograph is an effort by AMTE to provide further support for the work of mathematics teacher educators throughout the United States.

The members of AMTE are involved in different aspects of mathematics teacher education. Some focus on helping teachers, both preservice and inservice, develop a deep understanding of mathematics; others focus on helping teachers develop rich pedagogical content knowledge. Many work primarily with preservice teachers while others work mainly with inservice teachers. Additionally, some work within college or university classroom settings; others are "in the field" most of the time. Furthermore, for many mathematics teacher educators, responsibilities shift periodically.

In order to grow professionally as mathematics teacher educators, we participate in various professional organizations. The Mathematical Association of America, the National Council of Teachers of Mathematics (NCTM), the National Council of Supervisors of Mathematics, the American Mathematical Association of Two-Year Colleges, and the American Educational Research Association are just a few of the organizations in which we participate. However, mathematics teacher education is only one of many foci of these organizations.

When AMTE held its first annual meeting in Washington, DC, in 1997, it marked the beginning of a new professional development opportunity for mathematics teacher educators. At the AMTE meetings, all sessions focus on the day-to-day work of mathematics teacher educators. Moreover, it is possible to share and learn about different aspects of mathematics teacher education. For example, we can learn about the teaching of abstract algebra, integrating mathematics and science methods courses, and helping inservice teachers conduct a lesson study. Many mathematics teacher educators are excited about the opportunity to communicate and collaborate with their colleagues from across the country, and the AMTE annual meetings continue to flourish.

However, not all AMTE members can attend the AMTE annual meetings to take advantage of such a professional development opportunity. In addition, a number of mathematics teacher educators have lamented the lack of opportunity for sharing their own works more often, in particular through publications. The school-based journals of NCTM do publish some articles related to mathematics teacher education. Also, the *Journal of Mathematics Teacher Education* and the *Journal for Research in Mathematics Education* publish research articles related to teacher education. However, there has been no publication dedicated to the sharing of the day-to-day work of mathematics teacher educators. AMTE is the organization most suited to address this issue because it aims to facilitate and promote communication and collaboration among mathematics teacher educators.

In 2002, the AMTE Board decided to publish its first monograph and appointed the two of us as co-editors. This monograph was to be a "forum for mathematics teacher educators to exchange ideas about their work with preservice and inservice teachers and about their collaborative efforts with others who play significant roles in mathematics teacher education." Editorial panel members and a board liaison were also appointed. Together this group wrote a call for manuscripts that was published in February, 2003. Following the Board's recommendation, we also commissioned an anchor paper by Glenda Lappan.

Although as editors we felt this was a publication long needed, and many people communicated to us their agreement, we were not sure what level of response we would receive to the call for manuscripts. Because mathematics teacher educators are engaged in diverse aspects of teacher education, we needed quality manuscripts on a range of topics. As the deadline approached, the manuscripts started arriving, and they kept coming. We received 44 manuscripts, which provided a clear indication that many mathematics teacher educators were seeking opportunities to share their work with their colleagues across the country.

Three or four people reviewed all manuscripts. The co-editors and the board liaison met in Charlotte, NC, in October, 2003, to make some very difficult decisions based on the reviews. Although there were many quality manuscripts, page limitations forced us to accept only a few. After two days of deliberation, we accepted 12 manuscripts (27%).

We hope that readers will find this monograph to be useful in their day-to-day work with mathematics teachers. We also hope that this will be the first of many publications produced by AMTE. We know that learning to be a mathematics teacher is a life-long process, but so is learning to be a mathematics teacher educator. Just as mathematics teachers need opportunities for professional growth, so do mathematics teacher educators. Publications such as this monograph can serve an important role in the professional development of mathematics teacher educators.

We would like to thank AMTE for giving us the opportunity to work on this monograph. We would also like to express our sincere appreciation to the following individuals for their support in the production of this monograph.

Editorial Panel Members
    Susan Beal, St. Xavier University, Chicago
    Michaele F. Chappell, Middle Tennessee State University
    Dale Oliver, Humboldt State University, CA
Editorial Panel Board Liaison
    David Pugalee, University of North Carolina at Charlotte
AMTE Past-President
    Francis (Skip) Fennell, McDaniel College, Westminster, MD
AMTE President
    Karen Karp, University of Louisville
AMTE Executive Director
    Nadine Bezuk, San Diego State University

Finally, we would like to thank **all authors** who submitted their manuscripts for consideration.

Tad Watanabe
The Pennsylvania State University
[txw17@psu.edu]

Denisse R. Thompson
University of South Florida
[thompson@tempest.coedu.usf.edu]

Lappan, G. and Rivette, K.
AMTE Monograph 1
The Work of Mathematics Teacher Educators
©2004, pp. 1-17

1

Mathematics Teacher Education: At a Crossroad

Glenda Lappan
Kelly Rivette
Michigan State University

The work of mathematics teacher educators takes many forms – curriculum development, research, professional development, policy making, assessment, and evaluation. This paper uses the Interventions entry point from the RAND Study Cycle of Knowledge Production and the Improvement of Practice to consider the work of mathematics teacher educators related to preservice teacher education, professional development, and curriculum.

When we find ourselves at a crossroad, our instinct is to aim in the direction we want to go to get to the place we want to be. This thinking implies a clear choice of the right direction. Yet, for many reasons, not the least of which is the rapidly changing world in which we live, there is no such clear choice for mathematics teacher educators. We have reached a place in our profession where we recognize that the only defensible directional goal is to commit ourselves to successive approximations toward the nirvana where every teacher of mathematics has the right stuff and a fire in his or her belly to reach every child. Consequently, mathematics teacher educators need to build a profession that thrives on the challenge of examining, with curiosity, commitment, and vigor, the hard problems inherent in the teaching and learning of mathematics and the preparation and lifelong support of K-12 teachers. No organization is better positioned than the Association of Mathematics Teacher Educators (AMTE) to lead the development of such a profession.

Mathematics teacher educators work in many different ways — curriculum development, research, professional development, policy, and assessment and evaluation, among others. Wherever our work is situated, it is the contribution we make to the mathematical learning of K-12 teachers and their students that ultimately matters. In this paper, we focus on three areas of work in which AMTE members engage and suggest some promising future directions. To begin, we use a schematic from the RAND Mathematics Study Panel as an organizer for situating the areas of work.

The RAND Study Panel Report

The RAND Mathematics Study Panel report, *Mathematical Proficiency for All Students: Toward a Strategic Research and Development Program in Mathematics Education* (RAND Mathematics Study Panel, 2003), presents a provocative cycle of research, development, improved knowledge and practice, and evaluation. The cycle leads to new research and new development with the goal of producing new knowledge and understandings and improving practice.

As Figure 1 makes clear, one can enter this cycle of improvement at different places and professionals in mathematics teacher education do so. In this paper we choose the *Interventions* entry point with sections devoted to work in preservice teacher education, professional development, and curriculum.

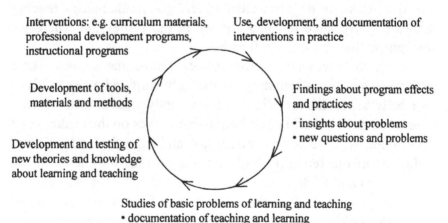

Figure 1. Cycle of Knowledge Production and the Improvement of Practice (Reprinted from *Mathematical Proficiency for All Students: Toward a Strategic Research and Development Program in Mathematics Education* with permission of the RAND Corporation ©2003. All rights reserved.)

The Preservice Education of Teachers

As a result of experience and research, mathematics teacher educators know that K-12 mathematics teaching involves a myriad of complexities beyond the organization of daily lesson plans, the development of test items, and classroom discipline. Although debates on what courses should be taught for preservice teachers, what the goals for these courses should be, and who should teach them remain, it is the responsibility of mathematics teacher educators to make suggestions and help make decisions about the directions to take. Whatever decisions are made, preservice teachers should exit their teacher preparation programs with experiences that have deepened their mathematical knowledge and helped them reflect on and defend or amend their own beliefs about the role of a mathematics teacher, mathematics as a discipline of study, and how mathematics can be effectively taught and learned.

The NCTM *Principles and Standards for School Mathematics* (2000) recommend that the different content strands of the mathematics curriculum be developed K-12: Number and Operations, Algebra, Geometry, Measurement, and Data Analysis and Probability. These content strands, along with the process strands of communication, reasoning and proof, problem solving, connections, and representation, can serve as a guide in the development of preservice teacher knowledge. The *Mathematical Education of Teachers* (MET) publication (Conference Board of the Mathematical Sciences, 2001) was created to stimulate improvement in programs for prospective teachers and offers recommendations on the nature of coursework for preservice K-12 teachers. MET includes classroom vignettes and examples to portray ways that teachers use their mathematical knowledge for teaching. The depth and breadth of mathematical knowledge described in the MET document includes the characteristics of understanding articulated by Liping Ma as a *profound understanding of fundamental mathematics content* (Ma, 1999). Ma defines profound understanding of fundamental mathematics as "an understanding of the terrain of fundamental mathematics that is deep, broad and thorough" (Ma, p.120).

In addition to the mathematical content offered in coursework, courses and field experiences can offer preservice teachers opportunities to reflect on the role of a mathematics teacher in helping students learn mathematics with deep, broad, and thorough understanding. Many of the papers in this monograph offer suggestions on engaging preservice teachers in reflection on the teaching of others and on their own teaching. Some suggestions include using alternative assessments to challenge mathematical beliefs (Coffey) and using audio- or videotape as a means of analyzing one's teaching (Taylor & O'Donnell). Using clinical interviews of children to assess teacher candidates in mathematics methods coursework can also challenge preservice teachers to examine their beliefs (Moyer-Packenham). Engaging preservice teachers in courses that "problematize familiar mathematical concepts" (Van Zoest, this volume, p. 121), require them to teach important concepts to a small group of middle-grades students, and then reflect on the experience with others can prompt an examination of beliefs about students' learning and what is important for students to know and be able to do.

Other ways to foster reflection on the role of a mathematics teacher are through Japanese Lesson study (see Bass, Usiskin, & Burrill, 2002) and videos of teachers in the field. Magdalene Lampert's and Deborah Ball's work around the video data they collected as part of the Mathematics And Teaching through Hypermedia (MATH) project at Michigan State University offers teacher educators a multimedia approach to mathematics teacher education (see Lampert and Ball, 1998). These examples of tools can give preservice teachers an opportunity to look into a classroom and discuss issues involving student and teacher knowledge, discourse, and decision-making.

Professional Development

Professional development activities are as old as schooling. However, the past two decades represent a time of particular focus in the U.S. on understanding and developing professional development strategies that have lasting payoff for teachers. In the papers in this monograph you will encounter a number of streams of work on professional development. We highlight a few promising directions in this paper by giving examples of professional development activities that focus on curriculum inquiry and students' mathematical work.

Curriculum Inquiry

Teachers, through their education and experience in the classroom, learn the skills and facts of the subject that they teach. There is often little in curriculum materials or the assessments for which teachers are held accountable that pushes toward conceptual rather than instrumental learning (Skemp, 1978). In our experience, it is extremely difficult for teachers to conceive of, plan for, and carry out teaching that has the goal of helping students make sense of ideas, connecting them to what they already know, and pushing further to see where they might lead, etc., without the support of good curriculum materials and opportunities for interaction with other teachers and professionals around that curriculum.

Curriculum materials can provoke dissatisfaction with limited learning opportunities for students. They can raise the possibility of engaging students with subject matter content in challenging ways. However, unless materials and professional development activities around the materials also provide opportunities for teachers to consider instructional strategies that support the goal of deeply understanding the concepts and related skills and procedures embedded in the materials, the curriculum does not reach its potential. As teachers enact curriculum they may or may not actually teach what the curriculum developers intend. So what can help teachers become classroom partners in curriculum development? One part of the answer is professional development. In 1999 and 2000, Janine Remillard published results from an interesting study and follow-up in which she examined what teachers learn from the enactment of reform-oriented text materials in their classrooms. In the two cases she developed, her study teachers learned mathematics and considered classroom instruction, but with very different levels of engagement and results. One indicator of these differences is shown in the following table (Remillard, 1999).

Table 1. Teachers' Patterns in Reading the Textbook During Task Selection

	Number of lessons observed	Number of lessons involving text	Number of lessons including item from...		
			Student page	Margin of student page in teacher's guide	Supplemental pages of teachers' guide
Jackie	14	4	3	4	4
Catherine	15	14	14	14	1

From her study of these two teachers, Remillard built a model to portray the interaction of her two subjects' arenas of influence while implementing the curriculum: Jackie and Catherine's beliefs about learning; their views of mathematics; and their interaction with the materials. Her model raises the issue of what professional development opportunities help teachers focus on the enactment of curriculum materials.

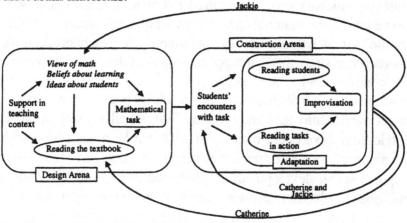

Figure 2. Model of the design and construction arenas in Catherine's and Jackie's teaching, illustrating the relationship between each arena and the influential factors within each

Remillard is a Co-PI of *Mathematics in America's Cities,* the newly funded Center for Teaching and Learning involving Rutgers, City College of New York (CUNY), and the University of Pennsylvania. The work of this center will be of interest to AMTE members as the Center's goal is to produce knowledge that can inform mathematics teaching practices in urban settings. All the districts involved have adopted new reform-oriented mathematics curriculum materials. Remillard's previous research suggests that professional development focused on the work of teachers in their classrooms as they implement new curricula can be successful in engaging teachers more deeply in examining their practice.

In the Classroom Coaching Study conducted from 1985 to 1987, our research and development group at Michigan State University studied the effects of classroom coaching on teachers' practice and their ability to adapt the teaching strategies they were learning from interaction with coaching and the Middle Grades Mathematics Materials (MGMP) (Fitzgerald, Lappan, Phillips, Shroyer, & Winter, 1986) to their traditional text. The 12 teachers in the study were assigned to three treatment groups. Four participated only in the intensive whole group professional development activities over the two years; four, in addition, had a classroom coach at three extended time periods over each year while teaching an MGMP unit; and four, in addition to classroom coaching, were coached to work with another teacher in his or her building in the second year (Fitzgerald, Lappan, Phillips, & Winter, NSF Final Report, NSF/MDR–83-18218). One goal of our work was to direct teacher's attention from student behavior to student cognition through our professional development and coaching.

Through analysis of the documentation, which included field notes from classroom observers, periodic surveys of teachers' beliefs about teaching and student learning, assessment of teachers' mathematical knowledge, and periodic teacher interviews, we found that coaching was effective in helping teachers improve their teaching and in shifting their primary focus to student cognition. However, we were surprised by our observations when one of the coached teachers and one of the un-coached teachers each had a student teacher. The un-coached teacher had no processes or effective language through which to analyze or plan lessons. She was a good teacher, but her classroom strategies were instinctive rather than the result of conversation and reflection. The coached teacher was highly effective in helping her student teacher. Through the previous year of periodic intensive analysis and planning of mathematics lessons together, she and her coach had developed a language with which to discuss classroom practices. This language helped her to be an effective coach of others. This mid-80's work supports the current push toward developing communities of practice within which to support teacher learning.

Student's Mathematical Work

In 2002, the Victoria Department of Education and Training, the Catholic Education Office (Melbourne), and the Independent Schools in Victoria issued a comprehensive report on the Early Numeracy Research Project (ENRP), currently available on CD-ROM. This report outlines the results of a large-scale study, directed by Doug M. Clarke (Clarke et al., 2002). As stated in the report (p. 11),

> The stated aims of the Early Numeracy Research Project were the following:
> - to assist schools to implement the key design elements as part of the school's mathematics program;
> - to challenge teachers to explore their beliefs and understandings about how children develop their understanding of mathematics, and how this can be supported through the teaching program; and
> - to evaluate the effect of the key design elements and the professional development program on student numeracy outcomes.

The study team developed a framework of "growth points" in young children's understanding of mathematics across a range of mathematical domains. Once the mathematical analyses were done and the framework was developed, the rest of the development work and the professional development activities were driven by progress on children's attainment of these "growth points."

Throughout the trial and reference schools at the three grades studied, 34,398 students were interviewed one-on-one by their teacher. The sheer scope and magnitude of the study is stunning as is the growth of the students in the trial sites over all mathematics domains tested. The growth of the trial teachers in their teaching effectiveness and content knowledge was equally impressive. The teachers' views changed from expecting the researchers to give them a recipe for what they should be doing to embracing the notion of creating *rich ingredients* for their practice that they combined to meet the needs of individual children.

Two of the key recommendations from this study show the intertwined nature of mathematical goals in curriculum inquiry and examining student work.

Recommendation 6:

> It is recommended that professional development provided to preservice and inservice teachers in early mathematics teaching and learning give particular focus to the ENRP Growth Points as a basis for guiding teachers' thinking and for assessment and planning.

Recommendation 7:

> It is recommended that the summary of the practices of effective mathematics teachers that emerged from the ENRP case studies, and the documented changed practices of trial school teachers form the basis of professional development on mathematics teaching practices. (p. 28)

The report makes clear that professional development that engages teachers in studying the progress of their students on well articulated goals and indicators of learning helps teachers improve their classroom practices and adds to their understanding and command of the mathematics they need for teaching. What we need now are a series of projects that systematically extend such analytic work and study to higher grades in K-12 mathematics education.

Research over the past decade points to the need for teachers to confront what their students know and do not know. This implies that students' work, whether it be test performance or regular class activity, can be a powerful tool for causing the dissonance needed to capture teachers' attention and curiosity. Current professional development studies in the U.S. share a common core of ideas about effective professional development:

- The classroom should be used as a laboratory for exploration of teaching and learning so that professional development interacts with teachers' regular teaching responsibilities.
- Teachers should be equal partners in improving mathematics teaching and learning.
- Video cases, written cases, and other artifacts of classroom practice, such as student work and assessments, can be used to promote deep analysis of teaching practices and of mathematics for teaching.
- Good curriculum materials chosen by the school can be an effective focus for professional development.

Curriculum

Jerome Bruner, in the 1990 Karplus address given at the annual National Science Teachers Association meeting, paid a tribute to Karplus (Bruner, 1992):

> What he knew was that science is not something that exists out there in nature, but that it is a tool in the mind of the knower—teacher and student alike. (p. 5)

Bruner goes on to say

> Getting to know something is an adventure in how to account for a great many things that you encounter in as simple and elegant a way as possible. And there are lots of ways of getting to that point, lots of different ways. And you don't really ever get there unless you do it, as a learner, on your own terms. All you can do for a learner enroute to their forming a view of their own view is to aid and abet them on their own voyage. (p. 5)

Although these remarks were made about science, they resonate with the reformulation of curriculum goals for mathematics K-12 worldwide.

For over three decades, a variety of national and international studies of student achievement have suggested that American practices in mathematics education are not yielding the kind of learning that is both desirable and possible (e.g., Second International Mathematics Study, National Assessment of Educational Progress, Third International Mathematics and Science Study, Programme for International Student Assessment). Comparing curricula and instructional practices in U. S. schools with those in countries with high student achievement revealed intriguing ideas about ways to improve our own results (e.g., Hiebert & Stigler, 2000; McKnight, et al.,1987; McKnight & Schmidt, 1998; Stigler, Lee, & Stevenson, 1990). Studies of mathematics curricula and teaching in Asian and European countries that are our intellectual and economic competitors showed that, in general,

- Our curricula do not challenge students to learn important topics in depth (Schmidt et al., 1999);
- Our teaching traditions encourage students to acquire routine procedural skills through a passive classroom routine of listening and practicing;
- Our assessment of student knowledge emphasizes multiple choice and short answer responses to low-level tasks. (McKnight et al., 1987).

Against this backdrop, the recommendations in the National Council of Teachers of Mathematics *Standards* documents (1989, 1991, 1995) called for major changes in traditional patterns of curriculum, teaching, and assessment in school mathematics. These ideas were refined and enhanced in the NCTM *Principles and Standards* in 2000. However, to make a difference, the vision of curriculum articulated in the *Standards* volumes as well as teaching, learning, and assessment around that curriculum has to be translated into effective and practical models.

In a paper presented at an AAAS symposium on curriculum development, Phillips, Lappan, Fey, and Friel (2001) articulated the following principles for curriculum materials that are emerging out of current curriculum development work.

- An effective curriculum has coherence—it builds and connects from investigation to investigation, unit to unit, and grade to grade.
- The key mathematics ideas around which the curriculum will be built are identified.
- Each key idea is related to a number of smaller concepts, skills, or procedures that are identified, elaborated, exemplified, and connected.
- Mathematics tasks that will form the work of students both inside and outside of class are the primary vehicle for students' engagement with the concepts to be learned.
- Posing mathematics tasks in context provides support for both making sense of the ideas and for processing them so that they can be recalled.
- Ideas are explored in sufficient depth to allow students to make sense of them.

These principles are more apparent in the new generation of mathematics curriculum materials that have emerged in the past decade. They represent a way of monitoring the development of new curricula as well as analyzing existing curriculum materials. They also suggest that the kind of curriculum development that has begun over the past decade represents demanding scholarship that needs to be recognized and rewarded in higher education faculty reviews. The work of figuring out how to assess the scholarship of curriculum development is a challenge that we as a field need to undertake.

The National Science Foundation, in recognition of the importance of curriculum in improving mathematics education, has recently funded two centers to focus on curriculum research: the Center for Curriculum Materials in Science (CCMS); and the Center for the Study of Mathematics Curriculum (CSMC). A sample of the kinds of questions that the mathematics curriculum center (Reys, Lappan, & Hirsch, 2003) will pursue follows.

Curriculum Design
- What principles for the design of curriculum materials can be drawn from contemporary research and theory on learning?
- How can mathematics curriculum materials be designed to serve the learning needs of students from diverse cultural

backgrounds and help to reduce the achievement gap among various populations of students?
- How might mathematics curriculum materials be designed to capitalize on continued advances in computing technologies?
- How can the knowledge gained from experienced mathematics curriculum developers be used to inform and prepare future writers of curriculum materials?

Curriculum Implementation
- What are the key factors involved in adopting and successfully implementing high quality mathematics curriculum materials?
- To what extent and under what conditions do mathematics curriculum investigation and implementation serve as vehicles for professional development and teacher learning?
- To what extent and under what conditions can mathematics curriculum materials promote teacher learning and effective teaching?

Curriculum Evaluation
- What are some of the critical features of mathematics curriculum materials that support student learning?
- What is the relationship among local or state curriculum standards, curriculum materials, and high stakes assessments?
- What evaluation tools are most effective when studying mathematics curriculum? What information do they yield for various purposes? What are their limitations?
- In what ways are evaluation data used to make school-based curriculum decisions?

CSMC has as an ongoing goal to develop tools and to inform and stimulate the field to engage in the study of curriculum and curriculum development as a vehicle for student and teacher learning. With systematic attention to the problem, perhaps in five to ten years we will be in a different place in our understanding of scholarship around the study of curriculum and be able to support young professionals with such interests through the higher education reviews that are a part of monitoring the quality of our work.

Summary

The *Cycle of Knowledge Production and the Improvement of Practice,* articulated by the RAND Study Panel chaired by Deborah Ball and illustrated in Figure 1, can be an important tool in examining where the field is and what AMTE as an organization can promote to improve the teaching and learning of mathematics. The Cycle makes clear the need for intensive study of current interventions with the goal of understanding program effects and determining new questions such interventions raise. The Cycle also promotes the need for mathematics teacher educators to engage in serious, long-term work to build theories that can guide new stages or iterations of both development and research programs. Such theory building is improved by "trips" through the stages of the cycle. Theories that are well articulated and shared with other professionals can also guide the development of research instruments, analytic strategies, and coding schemes for making sense of data.

We recognize that there is a human capacity problem associated with the enactment of such a cycle of improvement. To make a research-based cycle work, there have to be sufficient numbers of well-educated professionals to carry out the work. AMTE can play a major role in helping to develop young professionals to engage in systematic, articulated work to improve mathematics teaching and learning. The National Science Foundation, through its Centers for Teaching and Learning (CLT) program, has human capacity building as a primary goal. We would encourage AMTE to encourage promising teachers and strong undergraduate and master's students to take advantage of the support for doctoral work at one of the CLT Centers.

The *Cycle of Knowledge Production and the Improvement of Practice* will not in and of itself automatically advance the field. However, it can help us build a more coherent knowledge base on which to educate current and future teachers of mathematics. Whether one's work is situated in a department of teacher education, a department of mathematics, a school district or state office, or somewhere else in the system, the Cycle can help us articulate the primary focus of our work and see ourselves as a part of a more connected enterprise dedicated to improving the teaching and learning of mathematics in this country.

References

Bass, H. E., Usiskin, Z., & Burrill, G. E. (2002). *Studying classroom teaching as a medium for professional development. Proceedings of a U.S.-Japan workshop*. Washington, DC: National Academy Press.

Bruner, J. (1992). Science education and teachers: A Karplus lecture. *Journal of Science Education and Technology, 1*(1), 5-12.

Conference Board of the Mathematical Sciences. (2001). *The mathematical education of teachers: Part 2*. Washington, DC: Mathematical Association of America.

Clarke, D., Cheeseman, J., Gervasoni, A., Gronn, D., Horne, M., McDonough, A., Montgomery, P., Roche, A., Sullivan, P., Clarke, B., & Rowley, G. (2002). *Early numeracy research project*. Australian Catholic University and Monash University, commissioned by Department of Education, Employment and Training, the Catholic Education Office (Melbourne), and the Association of Independent Schools Victoria: 257.

Fitzgerald, W., Lappan, G., Phillips, E., Shroyer, J., & Winter, M. (1986). *Middle grades mathematics project*. Menlo Park, CA: Addison – Wesley.

Hiebert, J., & Stigler, J. W. (2000). A proposal for improving classroom teaching: Lessons from the TIMSS video study. *Elementary School Journal, 101*(1), 3-20.

Lampert, M., & Ball, D. L. (1998). *Teaching, multimedia, and mathematics: Investigations of real practice*. New York: Teacher's College Press.

Ma, L. (1999) *Knowing and teaching elementary school mathematics: Teachers' understanding of fundamental mathematics in China and the United States*. Mahwah, NJ: Erlbaum.

McKnight, C. C., Crosswhite, F. J., Dossey, J. A., Swafford, S. O., Kifer, E., Travers, K. J., & Cooney, T. J. (1987). *The underachieving curriculum: Assessing U.S. school mathematics from an international perspective. A national report on the Second International Mathematics Study*. Illinois University, Urbana Dept of Secondary Education: International Association for the Evaluation of Educational Achievement. Champaign, IL: Stipes Publishing Company.

McKnight, C. C., & Schmidt, W. H. (1998). Facing facts in U.S. science and mathematics education: Where we stand, where we want to go. *Journal of Science Education and Technology, 7*(1), 57-76.

National Council of Teachers of Mathematics. (1989). *Curriculum and evaluation standards for school mathematics*. Reston, VA: Author.

National Council of Teachers of Mathematics. (1991). *Professional standards for teaching mathematics*. Reston, VA: Author.

National Council of Teachers of Mathematics. (1995). *Assessment standards for school mathematics*. Reston, VA: Author.

National Council of Teachers of Mathematics. (2000). *Principles and standards for school mathematics*. Reston, VA: Author.

Phillips, E., Lappan, G., Fey, J., & Friel, S. (2001). Developing coherent, high quality curricula: The case of the Connected Mathematics Project. Paper presented at the AAAS Symposium, Washington, DC.

RAND Mathematics Study Panel. (2003). *Mathematical proficiency for all students: toward a strategic research and development program in mathematics education*. Santa Monica, CA: RAND.

Remillard, J. T. (1999). Curriculum materials in mathematics education reform: A framework for examining teachers' curriculum development. *Curriculum Inquiry, 29*(3), 315-342.

Remillard, J. T. (2000). Can curriculum materials support teachers' learning? *Elementary School Journal, 100*(4), 331-350.

Reys, B., Lappan, G., & Hirsch, C. (2003). *Center for the Study of Mathematics Curriculum*. (http://mathcurriculumcenter.org/crq.htm)

Schmidt, W. H., McKnight, C. C., Cogan, L. S., Jakwerth, P. M., & Houang, R. T. (1999). *Facing the consequences: Using TIMSS for a closer look at U.S. mathematics and science education*. Boston: Kluwer.

Skemp, R. R. (1978). Relational understanding and instrumental understanding. *Arithmetic Teacher, 26*(3), 9-15.

Stigler, J. W., Lee, S. Y., & Stevenson, H. (1990). *Mathematical knowledge: Mathematical knowledge of Japanese, Chinese, and American elementary school children*. Reston, VA: National Council of Teachers of Mathematics.

Glenda Lappan, University Distinguished Professor, Department of Mathematics, Michigan State University, received her Ed.D. in mathematics and education from the University of Georgia in 1965. From 1989–91 she served as the Program Director for Teacher Preparation at the National Science Foundation. From 1998–2000 she served as President of the National Council of Teachers of Mathematics (NCTM). Her research and development interests are in the connected areas of students' learning of mathematics and mathematics teacher professional growth and change at the middle and secondary levels. She has published over a hundred scholarly papers and numerous books for middle-grades students and teachers. She served as the Chair of the grades 5–8 writing group for the *Curriculum and Evaluation Standards for School Mathematics* (NCTM, 1989), and as Chair of the Commission that developed the *Professional Standards for Teaching Mathematics* (NCTM, 1991). She served as President of NCTM during the development and release of the 2000 *NCTM Principles and Standards for School Mathematics*. She is currently the Director of the Connected Mathematics Project II, which was funded by the National Science Foundation to develop a second iteration of Connected Mathematics Project I, a complete middle-school mathematics curriculum for teachers and for students. In addition, she is currently Chair of the Conference Board of the Mathematical Sciences and Vice Chair of the U.S. National Commission on Mathematics Instruction. [glappan@math.msu.edu]

Kelly Rivette, graduate student in mathematics education at Michigan State University, is interested in curriculum development. For the past two years she has worked as a research assistant for the Connected Mathematics Project II, a National Science Foundation funded middle-school curriculum. In the future, she hopes to teach mathematics content courses to preservice and inservice teachers. [rivettek@mail.msu.edu]

Tarr, J. E. and Papick, I. J.
AMTE Monograph 1
The Work of Mathematics Teacher Educators
©2004, pp. 19-34

2

Collaborative Efforts to Improve the Mathematical Preparation of Middle Grades Mathematics Teachers: Connecting Middle School and College Mathematics[1]

James E. Tarr
Ira J. Papick
University of Missouri-Columbia

In a collaborative effort funded by the National Science Foundation, mathematicians and mathematics teacher educators from the University of Missouri sought to improve the mathematical preparation of middle-grades teachers by developing four foundational courses in mathematics and supporting materials. The courses provide a strong mathematical foundation by connecting college-level mathematics with the mathematics taught in middle schools. Standards-based middle-grades mathematics curricular materials serve as a launching pad to explore and learn mathematics in greater depth. We report the nature of our courses, including specific conceptual examples, and provide a discussion of strategies for overcoming impediments to cross-college collaboration.

[1] This work was partially funded by a grant from the National Science Foundation (# ESI 0101822). The findings and opinions expressed are those of the authors and do not necessarily reflect either the position or policy of the National Science Foundation. We are grateful to our colleagues David Barker, John Beem, Joe Cavanaugh, Terry Goodman, Asma Harcharras, Dorina Mitrea, Debbie Perkowski, Mike Perkowski, Barbara Reys, and Robert Reys for their assistance in the development of this paper.

Over a decade ago, classroom mathematics teachers, mathematics teacher educators, and mathematicians combined efforts to create the foundations for *Standards*-based reform (Leitzel, 1991; National Council of Teachers of Mathematics [NCTM], 1989; National Research Council [NRC], 1989). In the early 1990s, the National Science Foundation (NSF) sought to improve K-12 student learning by supporting the development of several K-12 mathematics curriculum projects (*Standards*-based curricula), including five middle-grades curricula: *Connected Mathematics Project* (Lappan, Fey, Fitzgerald, Friel, & Phillips, 1998a), *Mathematics in Context* (National Center for Research in Mathematical Sciences Education & Freudenthal Institute, 1997-1998), *MathScape* (Education Development Center, Inc., 1998), *MathThematics* (Billstein & Williamson, 1999), and *Pathways to Algebra and Geometry* (Institute for Research on Learning, 2000). Comprehensive in nature, these *Standards*-based curricula present mathematics differently and in greater depth than traditional texts. They also introduce many important topics that have traditionally been reserved for the high-school level (e.g., number theory, conditional probability) through carefully-designed explorations that integrate concepts in algebra, geometry, data analysis and probability, and the underpinnings of calculus (Star, Herbel-Eisenmann, & Smith, 2000). These distinguishing characteristics of *Standards*-based curricula have created new challenges for teacher education.

The Need to Strengthen the
Mathematical Preparation of Teachers

Concurrent with the development of *Standards*-based mathematics curricula, the Missouri Department of Elementary and Secondary Education created teaching licensure for middle-school (5-9) mathematics. This event resulted in state-level discussions about middle-school mathematics teacher preparation and brought together mathematicians and mathematics teacher educators at the University of Missouri who subsequently collaborated on the NSF-funded *Missouri Middle Mathematics Project* (*M³ Project,* 1995-98) and the *Show-Me Project* (1997-2001, 2002-2005). This group of mathematicians and mathematics teacher educators were unified by their beliefs that middle-school mathematics teacher qualifications (both content and pedagogy) need to be strengthened in order to implement these new curricular materials successfully (Papick, Beem, Reys, & Reys, 1999; Reys, Reys, Barnes, Beem & Papick, 1997; Reys, Reys, Beem, & Papick, 1999). These sentiments are echoed by the authors of the National Research Council publication, *Educating Teachers of Science, Mathematics, and Technology* (2002), who cite results from a variety of teacher licensing examinations as compelling evidence that many K-8 teachers do not have sufficient content knowledge or adequate background for teaching mathematics, traditional or reform-based.

The infusion of *Standards*-based curricular materials in middle-grades classrooms, coupled with recent recommendations (e.g., Conference Board of Mathematical Sciences [CBMS], 2001), have necessitated changes in the mathematical preparation of teachers. National documents (CBMS, 2001; NRC, 2002) advocate greater collaboration between mathematicians and mathematics teacher educators in developing new programs for the preparation of mathematics teachers. Accordingly, college-level foundational mathematics courses and accompanying materials are needed to introduce and explore significant mathematical concepts and directly relate them to *Standards*-based middle-school mathematics curricula. Acquiring mathematical knowledge along these lines will allow teachers to understand how the mathematics they are teaching connects to more sophisticated mathematical ideas.

Connecting Middle School and College Mathematics

In a collaborative effort funded by the NSF, mathematicians and statisticians within the College of Arts and Sciences and mathematics teacher educators within the College of Education at the University of Missouri have responded to calls for changes in the mathematical preparation of middle-school teachers. Our three-year project, *Connecting Middle School and College Mathematics* (www.teachmathmissouri.org), is developing four foundational mathematics courses and accompanying curricular materials that provide middle-grades mathematics teachers with a strong mathematics foundation and connect college mathematics with the mathematics they will be teaching. The courses focus on algebraic structures, geometric structures, data analysis and probability, and the mathematics of change (calculus) and employ *Standards*-based middle-grades mathematics curricular materials as a launching pad to explore and learn mathematics in depth. Courses and accompanying materials were piloted in Summer Institutes (2002 and 2003) and outreach posts (off-campus, school-based academic year courses, 2001-2004) across the state of Missouri and are currently being piloted at numerous universities in the United States.

Our newly-created content courses are distinctively targeted to preservice middle-grades mathematics teachers and are designed to: (a) incorporate recommendations of the NCTM and the CBMS; (b) broaden and deepen content and pedagogy connections with *Standards*-based middle-grades mathematics curricula; and (c) provide a core mathematics foundation for individuals seeking to obtain middle-grades mathematics certification. Moreover, our structural design is aligned with national content and pedagogy standards, and incorporates critical components from a variety of middle-grades *Standards*-based mathematics curricular materials. In implementation, we frequently structure the classroom environment to parallel the *Standards*-based classroom (Ball, 1996) by engaging undergraduates in collaborative and individual learning situations, and employing multiple forms of authentic assessment, including open-ended problem-solving tasks, group projects (e.g., exploratory data analysis), and peer lesson presentations.

Our curricular materials have been constructed as four individual Companion Modules that serve as *Standards*-based supplements to college texts traditionally used in related courses (e.g., Banks, 1994; Bennett, Briggs, & Triola, 2000). Depending on the course, approximately 40-75% of the course is supplemented or replaced by content from a companion module. Although many traditional college mathematics textbooks have outstanding mathematical content, they typically do not provide the preservice middle-grades mathematics teacher with critical connections to the mathematical content that they will someday teach. Moreover, university mathematicians who teach content courses for preservice teachers are typically unfamiliar with *Standards*-based school mathematics and are not well positioned to make these critical connections. Thus, our Companion Modules seek to provide college content faculty with rich materials containing explicit mathematical and pedagogical connections to a variety of *Standards*-based curricula.

Each module embodies the style and philosophy of *Standards*-based middle-grades mathematics curricula in which students are actively engaged through inquiry (observation, experimentation, exploration, analysis, conjecture, proof, and generalization) as they learn significant and relevant mathematics. In fact, specific content references to the NSF-funded mathematics curriculum projects serve as focal points for discussion and analysis. Mathematical concepts are introduced and explored in the context of "real world" situations and then abstracted to more general settings as preservice teachers examine the underlying structure of each problem and discuss how the mathematics unfolds in the investigation.

An Advisory Board of mathematicians, statisticians, classroom teachers, mathematics supervisors, and mathematics teacher educators who have a serious commitment to *Standards*-based mathematics worked closely with our course development team in the development and refinement of each set of curricular materials. In particular, the Advisory Board met annually over the course of our three-year project. Prior to each meeting, drafts of chapters were distributed to Advisory Board members, who critically examined them with regard to both course content and presentation. Working sessions for each of the four courses were offered to capitalize on members' collective insights. The diversity of our Advisory Board guaranteed input from varied perspectives to enhance the quality and transportability of the course materials to other institutions. Upon reflection, we strongly believe that the Working Sessions format – in which individual contributions and perspectives are valued – fostered a sense of harmony among the key players in mathematics teacher education. In short, a shared commitment was made to enhance the mathematical preparation of middle-grades teachers.

Illustrative Examples from the Materials

To illustrate the nature of the courses, specific examples are provided from two companion modules: *Algebra Connections* and *Connecting Data Analysis & Probability*. Sample pages and the Table of Contents of all four modules are accessible on the project's website. (www.teachmathmissouri.org/courses/courses.htm)

Algebra Connections. The Fundamental Theorem of Arithmetic (FTA) and related concepts, such as greatest common divisor (GCD) and least common multiple (LCM), are fundamental in algebra and number theory. Middle-grades mathematics teachers need a firm understanding of these concepts in mathematical contexts as well as in real-world settings.

A typical class period of our *Algebra Connections* course related to these topics begins with undergraduates working in small collaborative groups of three to four to examine some specific *Standards*-based middle-grades mathematics curricular materials dealing with these concepts. For example, the sixth grade unit *Prime Time* from the *Connected Mathematics Project* (*CMP*) (Lappan, Fey, Fitzgerald, Friel, & Phillips, 1998b) introduces and develops many interesting mathematical explorations involving prime numbers, factorization, GCD, and LCM, and provides a framework to facilitate the understanding of the power and utility of the FTA.

A particular lesson from *Prime Time* (Investigation 4.2, Looking at Locust Cycles, page 38) focuses on different life cycles of cicadas (commonly called locusts) and asks several questions related to these cycles. For instance, Problem 4.2 raises the following questions involving LCM: "(a) how many years pass between the years when both 13-year and 17-year locusts are out at the same time? Explain how you got your answer; (b) Suppose there were 12-year, 14-year, and 16-year locusts, and they all came out this year. How many years will it be before they all come out together? Explain how you got your answer." The discussion on cicadas proceeds with a follow-up extension with added complexity: "Suppose cicadas have predators with 2-year cycles. How often would 12-year locusts face their predators? Would life be better for 13-year locusts?" (p. 42).

In the process of investigating this CMP unit, preservice teachers not only experience for themselves the concepts they will be teaching, but also have multiple opportunities to connect and deepen their understanding of these concepts. In particular, collaborative groups develop and test conjectures related to the relationship between the GCD and LCM using a variety of strategies. Subsequently, they share their findings with the whole class, and evaluate mathematical arguments presented by the groups. It is then convenient to consider generalities and computational tools for analyzing similar problems in more varied situations. For instance, the following collection of ideas naturally links to the mathematics encountered in this specific middle-school unit: (a) expressing the GCD and LCM of two integers in terms of the prime factors of the given integers; (b) determining the relationship between the GCD and LCM of two integers; (c) using and proving Euclid's algorithm to compute the GCD of two arbitrary integers (and the LCM using the discovered relationship); (d) employing the computations of Euclid's algorithm to write the GCD of two integers as a linear combination of the two integers; (e) and deriving several elementary number theoretic facts by exploiting the linear combination identity. The progression of these ideas occurs over several class periods and is integrated into other points of study throughout the course.

The power and versatility of the symbolic algebra/graphing calculator is fully exploited throughout the course, not only in computational venues, but in mathematical situations involving exploration, conjecture, discovery, and the construction of numerical or algebraic proofs. Questions regarding how the calculator performs its tasks lead to important discussions on the nature of algorithms and help provide a fundamental understanding of the role of technology in teaching and learning important mathematics.

Connecting Data Analysis & Probability. The authors of the CBMS report, *The Mathematical Preparation of Teachers*, assert that "of all the mathematical topics now appearing in middle grades curricula, teachers are least prepared to teach statistics and probability... [because] many prospective teachers have not encountered the fundamental ideas of modern statistics in their own

K-12 mathematics courses, and in fact need convincing that they need to learn this mathematics to be prepared to teach in the middle grades" (2001, p. 114). Consistent with this notion, preservice teachers entering the *Connecting Data Analysis & Probability* course were profoundly deficient in such content knowledge. Indeed, they would have benefited from a school mathematics experience consistent with the vision in *Principles and Standards for School Mathematics* (NCTM, 2000) which advocates a middle-school mathematics curriculum in which students formulate key questions, collect and organize data, represent data in a variety of ways, draw inferences from the data, and communicate their findings in a convincing manner. *Standards*-based curricula used in the course provided robust opportunities for preservice teachers to learn significant mathematics using a problem-solving approach to statistical ideas.

Preservice teachers in *Connecting Data Analysis & Probability* regularly work in collaborative groups, examining specific *Standards*-based middle-grades curricular materials dealing with key concepts. For example, the eighth-grade unit *Samples and Populations* from the *Connected Mathematics Project* (Lappan, Fey, Fitzgerald, Friel, & Phillips, 1998c) illustrates how statistical concepts can be applied in real-world contexts. This curricular unit explores many of the "big ideas" in data analysis and demonstrates the interrelationship between statistical concepts, displays of visual data, and techniques of data analysis. Moreover, this unit – as well as others in this series – models the NCTM's vision of a coherent school mathematics curriculum by offering connections between key statistical topics.

A particular lesson from *Samples and Populations* (Investigation 1, Comparing Quality Ratings, p. 23a) focuses on results of a consumer product study. More specifically, the data consists of information about the quality, sodium content, and price of 37 brands of peanut butter classified by four attributes: natural or regular, creamy or chunky, salted or unsalted, and name brand or store brand. Peanut butter is a common food in many households; with the increasing attention being paid to healthful diets, such nutritional information is important. In order to make informed purchasing decisions, numerous questions arise from the data. Is there a lot of salt in peanut butter? Is there much variation in quality ratings among different kinds of peanut butter? What is the best buy if I am most interested in quality? What is the best buy if I am interested in price? Preservice teachers use data from *Consumer Reports* to determine the existence of relationships between sodium content and quality rating, and investigate whether name brands of peanut butter outscore store brands in quality ratings. They use multiple displays of visual data and numerous data analysis techniques to justify their conclusions in order to convince their instructor and classmates.

Through their interaction with the *Standards*-based curricular materials, preservice teachers learn valuable connections between topics in data analysis and probability, as well as the importance of statistical processes as a means of solving real-world problems. After collaborative groups justify their inferences in a whole-class setting, they examine the topics in greater depth and with more mathematical rigor. In particular, investigation into whether a relationship exists between name brand/store brand and quality ratings gives rise to the study of non-parametric statistics. Examination of the relationship between sodium content and quality ratings leads to the formal study of correlation and regression models. Linear regression affords opportunities to introduce related statistical topics such as Mean Square Error, Pearson Correlation Coefficient, and the Method of Least Squares for determining a line of best fit. Moreover, the real-world contexts provide meaning for slope and vertical intercept. Not only do linear regression models serve as a powerful means of making predictions, they also provide a solid foundation for the study of proportionality as it relates to the coefficient for slope, further connecting the content to middle-school mathematics.

Graphing calculators are used extensively throughout the course, especially in the study of visual displays of data. In particular, data for the 37 peanut butters are entered as arrays for subsequent analysis. Statistical displays are created and analyzed. For example, quality ratings for regular brands can be displayed as a stem plot, rotated 90 degrees, and transformed into a histogram. Quality ratings for natural and regular brands are displayed as a back-to-back stem-and-leaf plot, and measures of central tendency are calculated. In addition, graphing calculators can readily display a variety of statistical plots (e.g., box-

and-whiskers plots, histograms, etc.). Working in groups, preservice teachers draw inferences – valid or invalid – from such visual displays of data, which serve as the focus of small group and whole-class discussions. Analysis and extensions of *Standards*-based investigations enhance preservice teachers' knowledge of concepts and processes in probability and statistics, and thus, position these future teachers to facilitate mathematical discourse and deliver instruction that promotes middle-school students' genuine understanding of the content.

Reflections on the Courses and Curricular Materials

Student Reflections. Feedback on our courses and curricular materials solicited by the project evaluator has been overwhelmingly positive. Teachers who participated in Summer Institutes and courses offered at outreach-posts consistently indicated that the courses helped them further develop their own mathematics understanding and provided them with ideas, instructional strategies, and resources that they could use with their students. In particular, they viewed what they were learning as being connected to and supportive of what they would be teaching to middle-grades students. Moreover, they felt that the Companion Modules provided (a) a strong mathematical foundation for a diverse audience of learners, (b) important connections between the mathematics being learned and the middle-grades curriculum, (c) connections among mathematics content strands as well as connections to other disciplines, (d) opportunities to engage in activities typical of a *Standards*-based learning environment, and (e) opportunities to develop an understanding of mathematics concepts through investigations, explorations, inquiry, proof, and generalization.

Teachers were impressed by the learning environment in each of the courses. Not only did it facilitate their learning, but it also provided them with a model to use in their classrooms. Teachers viewed the use of *Standards*-based curriculum materials and activities as being an important feature of the courses. These materials provided motivation to deepen their mathematics understanding, while at the same time providing them with rich examples of student-centered investigations and explorations appropriate for middle-school students.

Many teachers expressed some concerns about their lack of content knowledge; in some cases, these courses had helped to make them aware of this need. Reflecting about their own learning also helped them to think carefully about students' learning. Virtually all teachers, however, indicated that they were pleased with what they had learned and proud of the work they had done.

Faculty Reflections. Without reservation, we feel that our work on this project has been rewarding. Collaborative efforts between mathematicians, statisticians, and mathematics teacher educators have yielded innovative foundational mathematics courses for preservice teachers that blend content and pedagogy. Our collaborations have provided us with greater understanding of one another's perspectives on issues regarding the teaching and learning of mathematics. Moreover, there are several "lessons learned" that we feel compelled to share.

Early in the development of our Companion Modules, it became clear that *Standards*-based curricular materials offer abundant opportunities to study important, rigorous college mathematics. Stated simply, they are latent with powerful mathematical ideas that offer natural extensions to college mathematics. Additionally, in implementing the curricular materials we learned two important lessons. First, middle-school investigations used in launching the study of more sophisticated mathematics effectively addressed deficiencies in preservice teachers' content knowledge of school mathematics. In short, having not experienced a *Standards*-based school mathematics curriculum, all teachers benefited from these investigations. As one preservice teacher wrote in her journal, "the traditional curriculum is a far cry from this type of rigor!" Second, interactions with middle-school curricular materials prepared preservice teachers to implement *Standards*-based curricula more effectively in the future. As one teacher remarked, "I have learned so many different things but I feel the most important thing is how to use this information with my students… this course has encouraged me to emphasize higher-level thinking skills. I hope to use these ideas." Many preservice teachers indicated their beliefs about mathematics education were influenced by participation in our courses.

Strategies for Overcoming the Impediments
to Cross-College Collaboration

Given the impending teacher shortages, particularly in mathematics, universities across the nation are devoting more attention and resources to teacher education. Recent trends call for collaborative efforts between departments and colleges within universities to address this problem. Despite an increased emphasis placed on cooperation between departments and colleges, there are nonetheless impediments to collaborative efforts. Challenges related to this collaborative project are identified here, as are strategies for overcoming them.

Considerations for promotion and tenure as well as performance-based merit pay increases often differ significantly for faculty employed at institutions designated as Research I by the Association of American Universities than for those from institutions without such designation. More specifically, administration (e.g., department chairs, deans) at Research I institutions typically place more value on research papers published in refereed scholarly journals than on the development of curricular materials and service projects. The relative value placed on curriculum development is further differentiated between a Department of Mathematics within a College of Arts and Sciences and a Department of Learning, Teaching & Curriculum within a College of Education. Not surprisingly, the former offers less reward even though courses in development focus on the mathematical preparation of teachers. In addition, mathematicians who fail to perceive fault in traditional college mathematics courses devalue collaborative reform efforts and are typically unaware of *Standards*-based curricula or perceive them to represent too radical of a departure from tradition.

We have learned several effective strategies for promoting collaboration. At the national level, collaborative presentations at the Joint Meetings of the Mathematical Association of America (MAA) and the American Mathematical Society (AMS) serve to bridge divisions between mathematicians and mathematics teacher educators. Additionally, in 2004 our project will host a national conference on the mathematical preparation of teachers, engaging both sets of key players in discussion of this important component of teacher education. At the state level, the piloting of our materials by mathematicians at numerous state universities has heightened awareness of cross-college collaboration, and two state conferences on mathematics teacher preparation are planned. At our local level, colloquia have brought mathematicians and mathematics teacher educators together to discuss key issues regarding the mathematical preparation of teachers. Finally, the procurement of high-visibility, externally-funded projects has garnered the attention of faculty from both colleges and demonstrates what unified efforts can yield.

Related to the inter-departmental, cross-college nature of this project is the issue of governance. Because the courses in development will eventually be offered as upper-level courses in the College of Arts and Sciences, faculty from the College of Education may not be permitted to teach them under current policy of the Department of Mathematics. Thus, despite co-authorship and joint development of courses by faculty from both colleges, implementation and ongoing modification of the courses may ultimately be beyond the control of mathematics teacher educators. These policies are being examined and may be modified as a result of our project's success. Finally, even if policies were modified to enable mathematics teacher educators to teach for the Department of Mathematics, the College of Education would in essence be donating the faculty member's time to do so. In other words, each mathematics course staffed by a mathematics educator results in one less course that can be staffed in the College of Education unless deans from both colleges can settle on a compensatory package. Presently, our deans have managed to find common ground, but these issues nevertheless represent impediments to cross-college collaborative projects such as the one we describe here.

Conclusion

Notwithstanding the challenges facing our collaborative efforts, we believe this NSF-funded project holds the promise of improving the mathematical preparation of middle-grades teachers. A fundamental tenet of our courses and material development is that prospective middle-grades mathematics teachers should not only learn important mathematics, but they should also explicitly see the fundamental connections between what they are learning and what they will someday be teaching. Moreover, while learning this mathematics, they should directly experience exemplary classroom practice, creative applications to a wide variety of state-of-the-art technologies, and multiple forms of assessment, including open-ended tasks, group projects, and lesson presentations.

The introduction of *Standards*-based curricular materials poses new challenges for teacher education. Working as members of one team, mathematicians, statisticians, and mathematics teacher educators have collaborated to address these challenges and to improve the mathematical preparation of teachers.

References

Ball, D. L. (1996). Teacher learning and the mathematics reforms: What we know and what we need to learn. *Phi Delta Kappan, 77*(7), 500-508.

Banks, B. W. (1994). *A primer for modern mathematics.* Dubuque, IA: W. C. Brown Publishers.

Bennett, J. O., Briggs, W. L., & Triola, M. F. (2000). *Statistical reasoning for everyday life.* Boston, MA: Addison-Wesley.

Billstein, R., & Williamson, J. (1999). *Middle grades Math Thematics* (Books 1-3). Evanston, IL: McDougal Littell.

Conference Board of Mathematical Sciences. (2001). *The mathematical education of teachers.* Washington, DC: Mathematical Association of America.

Education Development Center, Inc. (1998). *Mathscape: Seeing and thinking mathematically* (6-8). New York: Glencoe/McGraw Hill.

Institute for Research on Learning. (2000). *Pathways to algebra and geometry.* Dallas, TX: Voyager Expanded Learning.

Lappan, G., Fey, J. T., Fitzgerald, W., Friel, S. N., & Phillips, E. D. (1998a). *Connected mathematics series*. Palo Alto, CA: Dale Seymour.

Lappan, G., Fey, J. T., Fitzgerald, W., Friel, S. N., & Phillips, E. D. (1998b). *Prime time*. Palo Alto, CA: Dale Seymour.

Lappan, G., Fey, J. T., Fitzgerald, W., Friel, S. N., & Phillips, E. D. (1998c). *Samples and populations*. Palo Alto, CA: Dale Seymour.

Leitzel, J. (Ed.) (1991). *A call for change: Recommendations for the mathematical preparation of teachers of mathematics*. Washington, DC: Mathematical Association of America.

National Center for Research in Mathematical Sciences Education & Freudenthal Institute. (Eds.). (1997-1998). *Mathematics in context*. Chicago: Encyclopedia Britannica.

National Council of Teachers of Mathematics. (1989). *Curriculum and evaluation standards for school mathematics*. Reston, VA: Author.

National Council of Teachers of Mathematics. (2000). *Principles and standards for school mathematics*. Reston, VA: Author.

National Research Council. (1989). *Everybody counts*. Washington, DC: National Academy Press.

National Research Council. (2002). *Educating teachers of science, mathematics, and technology: New practices for the new millennium*. Washington, DC: National Academy Press.

Papick, I., Beem, J., Reys, B., & Reys, R. (1999). Impact of the Missouri Middle Mathematics Project on the preparation of prospective middle school teachers. *Journal of Mathematics Teacher Education, 2*, 301-310.

Reys, B., Reys, R., Barnes, D., Beem, J., & Papick, I. (1997). Collaborative curriculum review as a vehicle for teacher enhancement and mathematics curriculum reform. *School Science and Mathematics, 97*(5), 250-256.

Reys, B., Reys, R., Beem, J., & Papick, I. (1999). The Missouri Middle Mathematics (M^3) Project: Stimulating *Standards*-based reform. *Journal of Mathematics Teacher Education, 2*, 215-222.

Star, J. R., Herbel-Eisenmann, B. A., & Smith III, J. P. (2000). Algebraic concepts: What's really new in new curricula? *Mathematics Teaching in the Middle School, 5* (7), 446-451.

James E. Tarr, Assistant Professor of Mathematics Education at the University of Missouri-Columbia, received his Ph.D. from Illinois State University in 1997. His research interests include middle-school students' probabilistic reasoning, the role of technology in teaching and learning mathematics, and the impact of *Standards*-based curricula on student learning. He was a member of the author team of the National Council of Teachers of Mathematics' *Navigating through Probability in Grades 6-8* (2003), and has collaborated with mathematicians and mathematics teacher educators at the University of Missouri on several large-scale National Science Foundation teacher enhancement projects. [TarrJ@missouri.edu]

Ira Papick, Professor of Mathematics at the University of Missouri-Columbia, received his Ph.D. from Rutgers University in 1975 in the field of Commutative Algebra. He has authored or co-authored research papers in that area, and is a co-author of the research book, *Prüfer Domains*, published by Marcel Dekker in 1997. Professor Papick has collaborated with mathematicians and mathematics teacher educators at the University of Missouri on several large-scale National Science Foundation teacher enhancement projects, and is currently the Principal Investigator of the three-year NSF project, "Connecting Middle School and College Mathematics." [papicki@missouri.edu]

Karp, A.
AMTE Monograph 1
The Work of Mathematics Teacher Educators
©2004, pp. 35-48

3

Conducting Research and Solving Problems: The Russian Experience of Inservice Training

Alexander Karp
Teachers College, Columbia University

This paper describes the inservice training experience in Russia. Using courses taught at a Russian university as an example, the author lays out the approach to organizing inservice training and describes the content of the courses. These courses devote considerable attention to solving difficult problems in school mathematics. The paper illustrates ways of posing interesting and substantive problems that generalize and develop ordinary school problems. It is argued that such courses would be useful in the United States in order to help teachers see the school course in mathematics not as a collection of rules, but as a living field of inquiry.

Many efforts have been made in recent years to determine the mathematics content knowledge that mathematics teachers must have (Ball, 1990, 1991; Conference Board of the Mathematical Sciences, 2001; Usiskin, Peressini, Marchisotto, & Stanley, 2003). Future teachers usually take a number of standard courses in mathematics, often the same courses taken by all those specializing in mathematics, whether future research mathematicians or teachers. Then, teachers go to a school where the subjects they teach turn out to be quite unlike what they have studied at the university. The need for intermediate courses of some type has been recognized for a long time (Klein, 1932), but the lack of such courses remains to this day. This paper describes the Russian experience of constructing courses for inservice teachers addressing this need for intermediate courses for high-school teachers.

Mathematics Education in Russia

Secondary Schools

Russian mathematics education has, especially during certain periods, attracted a great deal of attention from Western educators (e.g., McClintock, 1958; Swetz, 1978). Researchers have been particularly interested in schools for the mathematically gifted (Vogeli, 1968). At the same time, the actual practices of the Russian educational system, and the sources of its successes and failures, have remained largely unrecognized (Froumin, 1996).

In order to understand the construction of Russian mathematics teacher education, two principal characteristics of Russian education must be grasped. The first is the widespread involvement of professional mathematicians in secondary mathematics education (Vogeli, 1968). Professional mathematicians of the highest qualifications not only oversee the development of curricula, textbooks, and the organization of Olympiads and other competitions, but they also work directly with teachers and students. Sometimes they even become de facto directors of high schools. Consequently, professional mathematicians have exerted a considerable influence in determining the content and, to a great degree, the spirit of mathematics instruction. The second characteristic is the existence of mathematics entrance examinations in many colleges and universities (Swetz & Goldberg, 1976).

As a result of these two factors, school mathematics has evolved into a developed discipline with an arsenal of substantive problems. Consequently, the public has high expectations of teachers and their mathematics content knowledge. Good teachers are expected to prepare their students for their entrance examinations, which can often be demanding; it is certainly assumed that teachers will be able to solve the problems offered on these examinations. All of these factors together produce a favorable situation for the creation of special courses for teachers of mathematics.

Preservice Mathematics Teacher Education

Teacher preparation usually occurs in universities and pedagogical universities or colleges (Kerr, 1991). Traditional mathematics teacher preparation includes many advanced courses in mathematics: mathematical analysis, algebra (linear and abstract), number theory, geometry (multidimensional, differential, non-Euclidean), and the theory of probability. These courses are sometimes supplemented by other courses in mathematics, including courses in school mathematics (Curcio, Evans, & Plotkin, 1997). Methods courses often contain a substantial amount of specific mathematical material.

Inservice Mathematics Teacher Education

Russia (and the Soviet Union) had and continues to have a developed system of inservice education. This system usually operates within the framework of Teachers' Continuing Education Colleges (recently called Teachers' Colleges for Improving Qualifications, Universities or Academies for Post-Graduate Education, etc.). Naturally, inservice education is structured differently in different places. To speak about a general "Russian experience" is no easier than to speak about a general "American experience;" the differences between individual programs may be quite substantial. However, many of the ideas, themes, and approaches described in this paper from the work of the St. Petersburg Academy for Post-Graduate Education are popular in other Russian institutions (Karp & Nekrasov, 2000).

The St. Petersburg Academy for Post-Graduate Education offers various courses for teachers who have completed a preservice program and are working in schools. At the center of the program are the year-long courses, with classes held one day a week (six hours) for the entire school year (32-33 weeks). These courses are strongly recommended for all teachers every five years and are paid for by the government, which is the local educational authority. To make it possible for teachers to attend the courses, school administrators schedule classes so that the teachers attending the courses do not have any teaching responsibility on the day the course meets. Practically all mathematics teachers in St. Petersburg who have worked for an extended period of time have attended such courses.

The program usually consists of several sections. The psychological-philosophical section, which addresses issues such as conflict resolution, the specifics of adolescent psychology, familiarization with youth culture, etc., consists of about 34 hours. Approximately the same amount of time (32-34 hours) is devoted to general pedagogical issues, such as the relation between the family and the school, the organization of cooperative learning, innovative pedagogical methods, etc. Classes addressing pedagogical issues in mathematics are usually organized around presentations by invited specialists and consist of about 24 hours. Twelve hours are spent visiting schools, observing classes, and analyzing them. The remaining approximately 90 hours are devoted to the study of school mathematics.

The Academy also offers summer courses structured in a similar manner (with daily classes in June) and a variety of courses and workshops devoted to special issues. These special issues include courses for teachers working in specialized schools with an intensive program in mathematics, or courses for teachers working in schools that offer an abridged course in mathematics because of advanced teaching in the humanities, etc. To a certain degree, the materials used in these various courses overlap. For example, certain problems discussed in classes for teachers of intensive courses in mathematics and recommended for use with their high school students may also appear in the curricula of the year-long courses.

Courses in Mathematics

Purpose

Courses in mathematics are divided into courses in algebra, geometry, mathematical analysis, and finite mathematics. Because their contents are frequently altered and revised, the following description must necessarily limit itself to several isolated examples, without attempting to give a full account of the content of even a single course. Nevertheless, it should provide the readers a reasonable sense of how such a program is organized.

The courses are constructed with the assumption that coursework should lead to the following goals:

- mastery in mathematical skills, including facility with functional and set theoretic symbolism, algebraic and trigonometric transformations, construction of geometric figures in both two and three dimensions, use of various problem-solving methods (vectors, coordinates), etc.;
- familiarity with difficult problems, typical for a given course along with methods and techniques for solving them;
- the ability to use different means and forms of representation for presenting the same content and to make connections between different branches of mathematics;
- the development and implementation of logical thinking and abstract reasoning skills and the selection of examples to help students use their reasoning skills;
- openness to various approaches to solving problems, and the ability to pose new questions and problems (Zhigulev, 2002).

The issues on this list differ in character. In some cases, the aim is improvement – developing skills that were partly mastered in college and high school. In other cases, the aim is to broaden the teachers' knowledge, in particular, to acquaint them with new problems invented for Olympiads, entrance examinations, or classroom practice. In general, however, all of the courses are meant to engage teachers in specific mathematical activities, namely activities whose purpose is to help them acquire and develop characteristics of a "mathematical cast of mind" (Krutetskii, 1976). The organizers of these courses expect that this educational experience will subsequently be reflected in the teachers' own behavior in the classroom (e.g., a greater openness to new ideas and approaches).

Therefore, the courses are most often designed as sessions in solving problems that generalize and advance topics covered during ordinary high school lessons. The teachers taking these courses can, at times, make use of the materials presented to them in their own teaching; but the point of the courses is to attain a deeper understanding of what is missing from the textbook, and to develop the teachers' own mathematical reasoning abilities.

Mathematics Examples from the Courses

It is worthwhile to look at several examples of "advanced sections in school mathematics" and to examine the problems included in them.

a) *Solving equations and inequalities.* This section resembles 16th century algebra, one of whose important results was the derivation of formulas for solving cubic and fourth-degree equations. In order to solve difficult equations, teachers must think of ways to factor them or apply chunking techniques in which a whole expression is replaced by one variable. Such work requires facility in manipulating algebraic expressions, and often involves negotiating nontrivial logical questions, such as whether or not two equations are equivalent. In addition, in solving equations and inequalities, it is often useful to rely on connections with other fields of mathematics, such as analysis.

A single example will suffice. The equation $3^x + 4^x = 7^x$ can be solved in the following way. It is obvious that $x = 1$ is a root. The point of the problem is to show that no other roots exist. To show this, divide the left-hand and right-hand sides of the equation by 7^x and obtain $\left(\dfrac{3}{7}\right)^x + \left(\dfrac{4}{7}\right)^x = 1$. However, because the functions $y = \left(\dfrac{3}{7}\right)^x$ and $y = \left(\dfrac{4}{7}\right)^x$ are both strictly decreasing, the function $y = \left(\dfrac{3}{7}\right)^x + \left(\dfrac{4}{7}\right)^x$ is also strictly decreasing. Therefore, its graph cannot have more than one point of intersection with the horizontal line $y = 1$.

Although rarely used in Russia due to economic reasons, a graphing calculator would make the solution of this problem both clearer and easier to obtain. It would also involve teachers in a substantive discussion regarding the use of calculators: to what extent and under what circumstances can results obtained using a calculator be considered a rigorous proof? Thus, it seems fair to say that the application of technology would make such problems more fruitful.

b) *Elementary investigation of functions and the construction of graphs.* This section investigates functions without using calculus; again, a judicious use of calculators can be fruitful. The previous example made use of one result of such an investigation – that the sum of decreasing functions is itself a decreasing function. Other directions that elementary investigations of functions might take include the following:

- investigating how operations involving functions affect isolated properties (e.g., the sum of decreasing functions is itself a decreasing function; what can be said about their product?);
- investigating functions defined by formulas for monotonicity and convexity or determining the range of a function (usually reducible to proving an inequality);
- investigating the relationship between a formula that defines a function and the geometric transformations by which its graph can be obtained from a given graph (e.g., what formula defines a function whose graph is obtained from the parabola $y = x^2$ by the composition of a parallel translation by three units to the right along the x-axis and a reflection in the x-axis?);
- investigating the possibility of graphing a function with given properties and the definition of such a function by a formula (e.g., graphing a function whose domain is [0,1], and whose range is $[0,1] \cup [2,3]$).

c) *Problems with parameters.* This topic is remarkably popular in Russia. Simply taking the equation $x^2 - 4x + 3 = 0$ and replacing the number 3 with a parameter a to obtain $x^2 - 4x + a = 0$ makes it possible to pose many substantive questions. What number must a be in order for this equation to have real solutions? In order for it to have the solution $x = 1$? In order for it to have a solution greater than 3? This list of questions can be extended further. Solving such problems requires employing both graphic techniques and rather intricate reasoning. A particular advantage of problems with parameters is that these problems generalize ordinary school problems in a natural way. The reverse is true as well – the obtained answers can be checked by substituting the obtained values and turning a general problem into a specific one.

d) *Special points and lines in the triangle and the tetrahedron.* This section deals with topics close to those covered in the classic text of Coxeter and Greitzer (1967), along with analogous problems from three-dimensional geometry (Skopets & Ponarin, 1974). For example, after discussing that the three medians of a triangle intersect in one point, the following questions are posed:

- what segment in a tetrahedron can be considered to play the same role as a median in a triangle;
- what assertion could be formulated about such "medians;"
- how can such an assertion be proved?

In addition, whenever possible, questions are solved or proved in different ways. This demonstrates the significance of different techniques and leads teachers to discuss the possibility of using similar techniques in both two and three dimensions.

Common Features of the Mathematics Problems

All of the previous examples have certain common features. To characterize them, it is useful initially to list negative aspects often associated with the mathematical education of future teachers in the U.S.:

- The topics studied at the university level have little relation to actual practice at the high-school level.
- There is a prevalence, both at the university and high-school levels, of problems that Polya (1973) characterized as "routine." A "problem is a 'routine problem' if it can be solved either by substituting special data into a formerly solved general problem, or by following step by step, without any trace of originality, some well-worn conspicuous example" (p. 171).
- Classes are overloaded with new concepts. Virtually all class time is devoted to grasping and analyzing very basic examples of concepts, with almost no time for freely solving problems associated with them.
- There is a prevalence of problems meant only to illustrate the methods being described, and a corresponding reluctance to apply other methods in solving them.

The examples previously analyzed share none of these negative characteristics.

- All of the subjects discussed through the examples are connected with ordinary high-school mathematics and make use of ordinary high-school concepts; therefore, "transferring" logical and reasoning skills to the classroom situation is potentially easier than it is in the case of advanced courses.
- The number of new concepts introduced in these courses is limited a priori, apart from cases such as the last example in which the introduction of new concepts is itself an object of exploration.
- A variety of different methods are used in each case, and their application is either deliberately embedded in the course or is unavoidable because the problem lies at the intersection of several different content topics.
- Problems are usually solved in several steps. Although the transition from one step to the next is suggested and facilitated by experience, it is never rigidly dictated by a prescribed general routine. Additionally, problem sets often remain open in the sense that their construction often leads to new ways of posing the problems.

The difference between the two approaches described here can be characterized as a difference between an approach oriented toward research and an approach oriented toward repetition and reproduction. School mathematics contains many questions that are well suited to the former approach and are useful for further development of the subjects discussed during ordinary lessons.

Organization of Materials in Mathematics Courses

The format of the classes naturally depends on the inclinations of the instructors. There is usually a general discussion, conducted by the instructor, as well as individual problem-solving activities, sometimes done in pairs or small groups. The discussion reveals how, by continuing, developing, and generalizing standard school problems, it is possible to arrive at the more difficult problems being examined.

Problem solving is typically followed by a discussion of the ways in which the studied material is connected to the standard school curriculum. The issues that might be discussed include lesson construction related to the topics under investigation, typical mistakes that students make, and typical questions that students ask.

Solving difficult problems develops teachers' creativity and helps them to feel confident when they discuss related, but easier, materials with their students. For example, when students solve an equation, such as $x^2 - 4x + 3 = 0$, many of them experience difficulties because of an inability to monitor themselves and check their work; difficulties also arise when they confuse the different rules they have memorized. Thus, a student may obtain solutions of -1 and -3, and then do nothing to verify them. Another common problem arises from students' inability to represent the obtained results in different ways. After solving the equation $x^2 - 4x + 3 = 0$, a student may construct a graph of the function $y = x^2 - 4x + 3$, but get the points of intersection with the x-axis wrong.

Experience working with the equation $x^2 - 4x + a = 0$, which is one generalization of the equation $x^2 - 4x + 3 = 0$, may help teachers. Experiences acquired through the courses with using various forms of representation, and with investigating and posing new questions, enable the teacher to pose questions that will help students overcome the aforementioned difficulties. Teachers learn to combine problems in which students are asked to check the solution they have obtained in some way, or to compare it with another result. These problems are discussed in the courses for teachers, in terms of how they might be used in classes with high school students. Such collective discussions pave the way for further discussions of conducting more advanced lessons on solving equations with parameters. Other course topics are studied in an analogous fashion.

It is important to emphasize that the objects of investigation in the courses at the Academy are not individual problems, but problem sets. Problem sets usually begin with simple questions and gradually progress to more difficult ones, including questions that are considerably more difficult than the ones given as examples in this paper. The idea of using problem blocks is quite popular (Karp, 2002). In such cases, a single object (e.g., function, equation, geometric figure) is examined and questions are posed that allow this object to be examined from different angles. Usually such an investigation offers opportunities to prepare teachers for difficult problems and for making their solutions easier and trains them to search for interconnections, to reason by analogy, to generalize assertions, and to evolve other essential skills of intellectual activity.

Conclusion

Polya (1981) wrote:

> If, however, the teacher has had no experience in creative work of some sort, how will he be able to inspire, to lead, to help, or even to recognize the creative activity of his students? A teacher who acquired whatever he knows in mathematics purely receptively can hardly promote the active learning of his students. A teacher who never had a bright idea in his life will probably reprimand a student who has one instead of encouraging him (p.113).

Hardly anyone would argue with these conclusions. However, teachers are often not given any opportunity to acquire experience in creative work, sometimes because they are too busy completing a teacher education program and sometimes because it is assumed that teachers will not be able to handle creative work. The Russian experience described in this paper shows that teachers can be involved, en masse, in solving and discussing substantive mathematical problems of the kind that occur in large quantities in school mathematics.

The idea of creating similar courses for inservice and preservice teachers in the U.S. deserves some attention. Such courses differ in content from standard courses in mathematics (e.g., calculus, linear algebra, differential equations). They also differ from many courses devoted to general problem-solving principles and strategies, although they resemble some of the courses taught by Polya. In the proposed courses, strategies are illustrated by studying a single, sufficiently narrow topic; in the analysis of this topic, the formulations of problems are no less important than their solutions. Teacher-participants of these courses have something in common with research mathematicians whose work also focuses on the investigation of a specific area. A teacher who has completed such a course of study becomes accustomed to seeing objects for serious research in school mathematics.

Of course, it would be naive to consider everything in the Russian experience a total success. An outside observer who is not involved in the process of posing concrete problems sometimes finds them cumbersome and pointless. The absence of technology substantially limits the opportunities for research. It would obviously be inappropriate to speak of a direct transfer of the experiences of one country to another with a different culture and traditions. U.S. teacher educators must select and develop lesson topics and lesson formats while keeping in mind the specifics of the U.S. setting, although one would imagine that the approaches, themes, and problems used in Russia could likewise be useful here. In addition, U.S. teacher educators must be prepared to conduct such investigative work themselves; in this respect, collaboration with professional mathematicians in teaching courses and writing problems could be of great assistance.

Devoting some thought to the Russian experience may have fruitful consequences. The Russian experience demonstrates once again that school mathematics is a living field in mathematics. Familiarity with it might help teachers to bring into their classrooms a creative attitude toward the subject.

References

Ball, D. L. (1990). The mathematical understandings that prospective teachers bring to teacher education. *Elementary School Journal, 90,* 449-466.

Ball, D. L. (1991). Research on teaching mathematics: Making subject matter knowledge part of the equation. In J. Brophy (Ed.), *Advances in research on teaching,* vol. 2, 1-48. Greenwich, CT: JAI Press.

Conference Board of the Mathematical Sciences. (2001). *The mathematical education of teachers.* Washington, DC: Mathematical Association of America. (In cooperation with American Mathematical Society, Providence, Rhode Island.)

Coxeter, H. S. M., & Greitzer, S. L. (1967) *Geometry revisited.* Washington, DC: Mathematical Association of America.

Curcio, F. R., Evans, R. C., & Plotkin, A. (1997). Restructuring mathematics teacher education: The evolution of an innovative preservice program in Russia. *Teacher Education Quarterly*, Spring, 53-61.

Froumin, I. (1996). The challenge of Russian mathematics education: Does it still exist? *Focus on Learning Problems in Mathematics, 18*(4), 8-34.

Karp, A. (2002). Mathematics problems in blocks: How to write them and why. *PRIMUS, XII* (4), 289-304.

Karp, A., & Nekrasov, V. (2000). *Matematika v Sankt-Peterburgskoy shkole* (Mathematics in St. Petersburg Schools). St. Petersburg: Spezlit.

Kerr, S. T. (1991). Beyond dogma: Teacher education in the USSR. *Journal of Teacher Education, 42*, 332-349.

Klein, F. (1932). *Elementary mathematics from an advanced standpoint.* New York: Dover Publications.

Krutetskii, V. A. (1976). *The psychology of mathematical abilities in schoolchildren* (J. Kilpatrick & I. Wirszup (Eds.); J. Teller, Trans.). Chicago: University of Chicago Press.

McClintock, C. (1958). *The competition in education: US vs. USSR.* Santa Barbara, CA: General Electric.

Polya, G. (1973). *How to solve it: a new aspect of mathematical methods.* Princeton, NJ: Princeton University Press.

Polya, G. (1981). *Mathematical discovery: On understanding, learning, and teaching problem solving.* New York: Wiley.

Skopets, Z. A., & Ponarin, Ia. P. (1974). *Geometria tetraedra i ego elementov.* (Geometry of the tetrahedron and its elements). Iaroslavl: IaPI.

Swetz, F. (Ed.). (1978). *Socialist mathematics education.* Southampton, PA: Burgundy Press.

Swetz, F., & Goldberg, J. (1976). Mathematical examinations in the Soviet Union. *Mathematics Teacher, 70*, 201-218.

Usiskin, Z., Peressini, A. L., Marchisotto, E., & Stanley, D. (2003). *Mathematics for high school teachers: An advanced perspective.* Upper Saddle River, NJ: Prentice Hall.

Vogeli, B. R. (1968). *Soviet secondary schools for the mathematically talented.* Washington, DC: National Council of Teachers of Mathematics.

Zhigulev, L. (2002). Programmy godichnyh kursov (Year-Long Course Program), 2002-2003 (unpublished).

Endnotes

The author wishes to express his gratitude to L. Zhigulev and V. Nekrasov, Lecturers at the St. Petersburg Academy for Post-Graduate Education, for useful discussions and for sharing materials with him.

Alexander Karp, Assistant Professor of Mathematics Education at Teachers College, Columbia University, received his doctoral degree from Hertzen University in St. Petersburg, Russia. He taught middle school and high school for many years and has also taught preservice and inservice courses for teachers in various universities. Dr. Karp has authored a number of textbooks and collections of problems. His scholarly interests include teaching the mathematically gifted, teaching problem solving, and the history of mathematics education. [apk16@columbia.edu]

Coffey, D. C.
AMTE Monograph 1
The Work of Mathematics Teacher Educators
©2004, pp. 49-66

4

Using Alternative Assessment To Affect Preservice Elementary Teachers' Beliefs About Mathematics

David Charles Coffey
Grand Valley State University

This paper examines how the use of alternative assessments in a reform-based mathematics course affects preservice teachers' mathematical beliefs. A mathematics content course designed for elementary education majors employed three different alternative assessments to challenge counterproductive beliefs related to perseverance and predictability. Two of the three assessments are described, including the tasks and their rubrics, and samples of preservice teachers' reactions to these alternative assessments are discussed. Most preservice teachers in the course credited the assessments with changing their view of mathematics and anticipated that they would use alternative forms of mathematics assessments in their future classrooms.

Many preservice teachers perceive mathematics as a set of unchanging truths that exist separate from human experience (Brown & Borko, 1992; Dossey, 1992; Thompson, 1992). Teachers who hold this view typically believe that mathematics teaching involves disseminating content to students through lectures and demonstrations followed by structured student practice (Ball, 1990; Middleton, 1999; Schoenfeld, 1992). The perspective on mathematics and its instruction recommended by the National Council of Teachers of Mathematics (NCTM), however, stresses mathematical process as well as conceptual understanding of content (NCTM, 2000).

To prepare future teachers to teach from the perspective envisioned by NCTM, many mathematics educators design courses to challenge preservice teachers' view of mathematics as a pre-existing set of facts and procedures (predictability) that are used to solve some exercise quickly (lack of perseverance). Courses taught from this perspective usually engage preservice teachers in non-routine problem-solving situations that require them to construct mathematical understanding through reasoning and communication. Unfortunately, mathematical beliefs related to predictability and perseverance have proven extremely resistant to change (Pajares, 1992; Raymond & Santos, 1995). There is hope, however. Thus far most efforts have focused on addressing mathematical beliefs through changes to the content and instruction of mathematics education courses, not through changes in assessment practices (Bell, 1995).

Using traditional testing methods in courses meant to reflect reform-based mathematics conveys conflicting messages to preservice teachers. "As the curriculum changes, so must the tests. Tests also must change because they are one way of communicating what is important for students to know… In this way tests can effect change" (NCTM, 1989, pp. 189-190). These revised assessments must engage students in mathematical tasks that require them to construct and apply mathematical concepts in relatively new ways (Lambdin, Kehle, & Preston, 1996). Thus, assessment has the potential to challenge preservice teachers to rethink their traditional views by challenging their views with the piece of the educational system many value most – grades (Lester & Kroll, 1991; Wilson, 1994b). Consequently, I designed three alternative assessment tasks for a content course required for preservice elementary teachers to challenge these future teachers' mathematical beliefs.

Assessments and Preservice Teachers' Mathematical Beliefs

People construct beliefs about mathematics based on experiences (Cobb, Wood, & Yackel, 1990), and these mathematical beliefs, in turn, affect how a person acts during future encounters with mathematics (Pajares, 1992). The mathematical beliefs held by a majority of preservice teachers represent a narrow and static view of the subject, typically associated with traditional school mathematics (Dossey, 1992; McLeod, 1992; Raymond & Santos, 1995). Two of the most dominant mathematical beliefs found in the research literature concern predictability and perseverance in mathematics.

Schoenfeld (1988) offers a scenario that demonstrates how traditional mathematics tests might influence preservice teachers' beliefs regarding predictability and perseverance. He describes a unit test given in a high school class he observed which "contained 25 problems – giving students an average of 2 minutes and 10 seconds to work on each problem. The teacher's advice to the students summed things up in a nutshell: 'You'll have to know all your constructions cold so you don't spend a lot of time thinking about them'" (p. 159). Timed arithmetic tests that measure elementary students' computational ability offer another example of a commonly used assessment that rewards speed and accuracy in remembering pre-existing facts (Bell, 1995). The message seems quite clear – mathematics means determining a quick answer using a pre-existing, memorized method (Bell, 1995; Clarke, Clarke, & Lovitt, 1990; Hancock & Kilpatrick, 1993). Such tests fail to represent the true complexity of mathematics (Galbraith, 1993; Izard, 1993; Wheeler, 1993).

These mathematical beliefs are counterproductive because they interfere with the learning and teaching of mathematics as suggested by the NCTM (Ball, 1990; Cirulis, 1991). In order for success to occur, productive beliefs must replace the counterproductive ones. If students are to be successful in true mathematical problem-solving situations, then they must hold beliefs that support perseverance and flexibility in thinking (Schoenfeld, 1992). The assessments created for the probability and statistics course were intended to support the development of these more productive beliefs.

Assessments in Probability and Statistics

Probability and Statistics is a course required for preservice elementary teachers at Western Michigan University.[1] Teachers in the course work in groups of three or four and are expected to

---

[1] The results presented in this paper are part of the author's doctoral research completed at Western Michigan University. He continues to use alternative assessments similar to those described here at Grand Valley State University.

communicate their methods, solutions, and comprehension of the mathematics as they engage in non-routine mathematical tasks. The three alternative assessments, each worth roughly 20% of the final grade, reinforced this emphasis on collaboration, communication, and problem solving. Course participation (5%), weekly journals (5%), and a final exam (30%) completed the course grading.

The alternative assessment tasks addressed different topics from probability and statistics but were similar in structure. During each task, preservice teachers engaged in activities that teachers actually perform in their work, were encouraged to undertake the tasks in pairs, and were required to self-assess their work. The first two assessment tasks were the most authentic for the preservice teachers because they reflect actual activities of teachers – writing and evaluating tests. Thus, this paper focuses on two tasks, *Assessment Writing* and *Evaluating Responses*. (See Coffey 2000 for further information on these assessments and the third, *Monte Carlo Task*.)

Assessment Writing requires preservice teachers to write test questions to evaluate their peers' understanding of organizing and representing data. (See Appendix A for a description of each task and an accompanying rubric.) In pairs, preservice teachers worked on the task for over a month, including portions of two class periods. The pairs' assessment items were used to provide evidence of the depth of their understanding of the material.

Evaluating Responses, the second assessment task, directs new preservice teacher pairs to create an answer key for a pre-existing *Probability and Statistics* exam covering topics in descriptive statistics. Again, partners worked on their key during two class periods, but completed the majority of the task outside of class. Having developed an exam key, preservice teachers worked individually during one class period on the evaluation of a fictional individual's exam answers (see Figure 1 for a sample page). These answers represented typical responses on past exams covering the same material. Not only did the preservice teachers determine the correctness of each answer but they also identified misconceptions that might have resulted in a mistake and evaluated how a mistake ought to affect the score. For scoring, preservice teachers relied on the same rubric that the instructor used to evaluate their answer keys. Thus, *Evaluating Responses* gave the instructor an opportunity to assess preservice teachers' procedural understanding through the evaluation of the key and their conceptual understanding via their explanations of how they evaluated the fictional individual's exam.

On any two of the three assessment tasks, preservice teachers could correct their mistakes, thereby improving their understanding as well as their score. Although a small penalty, amounting to between 5% and 10% of the assessment grade, accompanied these second chance attempts, many individuals took advantage of the opportunity

---

MATH 265 FALL 1998 EXAM II

Name ___*Stew Dent*___

When responding to items on this test, please show all work, or steps taken, so that you will be eligible for partial credit.

1. The stem-and-leaf plot given at the left below shows the grade level reading scores of several high school students. Note that the **leaves have not been ordered** within each stem.

STEM	LEAVES
8	1
9	5 6 5 9 7
10	9 4 6 5 2 7 3 4 0 5 9 5 5
11	0 9 7 2 1 6 4 3 8 1 7 8 6 5
12	3 2 1 3

S	L
8	1
9	5 6 7 9
10	0|2 3 4 5 7 9
11	0 1 2 3 4 5 6|7 8 9
12	1 2 3

KEY: 10 | 3 REPRESENTS A GRADE LEVEL OF 10.3

   a. Determine the 5 critical numbers needed for constructing a box plot of these scores. **Name** and **give the value** of these 5 numbers. [10 pts]

   Name: *lower-extreme   lower-quartile   median   upper-quartile   upper-extreme*

   Value: ___*81*___   ___*101*___   ___*110*___   ___*116½*___   ___*123*___

   b. Draw a box-plot of the scores using the given line segment to define and label a scale for the plot. [5 pts]

   80   90   100   110   120   130

   c. Determine if any of the scores in this data set should be considered outliers. Justify your response. [5 pts]

   To be an outlier, the data value has to be more than 1.5 times the width of the box. 123 is no way an outlier. Too close to the upper quartile. Now 81 could be but I measured it out and it isn't an outlier.

Figure 1. First page of fictional work from *Evaluating Responses*
Note: Student responses (in italics) are not necessarily correct.

to resubmit their work or meet with the instructor to discuss their mistakes. Not only did these assessments emphasize that mathematics involves collaboration, connections to real-life, and self-reflection, they also demonstrated that mathematics is a process requiring perseverance.

Discussion

After completing both alternative assessments, preservice teachers were asked to write in their journals about the impact these assessments had on their beliefs about doing, learning, and teaching mathematics. The preservice teachers also discussed whether or not they would implement mathematics assessments like these in their future classes. Responses did not always address each issue, but they did provide insight into how these preservice teachers verbalized their beliefs regarding mathematics and alternative assessments.

Figure 2 offers a sample journal response. Although the journal structure included areas for four responses, most preservice teachers used the entire space to focus on the questions related to assessment and mathematics as the 'Prompt for the week.' The sample in Figure 2 also provides examples of perspectives articulated by many preservice teachers in their responses.

The sample response provides insight into the belief that mathematics is predictable with one method to get one answer: "Math should never be just memorizing. A student of course wants to get the right answer and remember it for the test but the math objective should be how to get to the answer and being able to explain to another person how to do the problem... These assessments have really made me believe what I have written above" (S1). This entry suggests recognition of the pitfalls associated with memorization but fails to address the learner's ability to invent effective methods to solve problems, a view typical among the preservice teachers.

> I think that math is both memorizing and constructing. To me they go together. The assessments made me see math in a totally new light. (S6)

Date: *11/5*

Prompt for the week: *I feel that these assesments* [sic] *gauge the students knowledge of a subject much better than a "normal" test. In my experience, when I have a "normal" test, I go over the material and what I don't know I spend extra time on but after the test I only feel relieved not still knowledgeable. However, with these assessments I spend time making up questions or evaluating work which requires me to* <u>understand</u> *the process.* ~~What questions do you have over any recent work?~~ *Math should never be just memorizing. A student of course wants to get the right answer and remember it for the test but the main objective should be how to get to the answer and being able to explain to another person how to do the problem. I unfortunately was never taught this way and I think this hinders my present math situation and possibly my future in teaching. These assessments have really made me believe what I have written above. I know now I need* ~~How is your group doing? How are you being a contributing member?~~ *to really stress concepts and processes as a teacher and if a student of mine understands this while doing a problem then most likely they will get the correct answer. When I began this class I was really frustrated with this method of learning. Now, however, I think all math and science classes should be taught like this one because I, unfortunately, do not think I remember much from math classes before this. Not only do I not remember much I am really enjoying this class.* ~~Anything else?~~

Figure 2. Sample preservice teacher journal entry (with prompts crossed out by student)

I think the assessments that we have done are a good idea because I know that it has forced me to have more of a thorough understanding of the material. (S2)

I really like the assessments. They're challenging me and making me look at the concepts in different ways – more completely I guess. (S7)

(T)hese assessments are a good tool for understanding. They make the students understand the concepts rather than just repeating procedures. (S3)

Giving students these kind [sic] of assessments can help students get a better understanding of how a problem is solved and explain why. (S4)

It's not simply about reading and taking tests. I actually get to use my brain, my hands, and my calculator. (S5)

Of the twenty-five responses, only the last one came close to suggesting that mathematics could be a truly dynamic and creative process.

In regards to perseverance, the preservice teachers were aware that the assessments required more time than a normal test. The extra time was evident in completing the initial task:

> ...I realized the huge difference in the # of hours we spent on the assessment vs. what I would usually spend studying for a test. (S8)

> ...there is more studying, research and work needed to do well. (S9)

Extra time was also needed because preservice teachers could improve their score by resubmitting their work:

> I really like the idea that we are able to redue [sic] problems. (S10)

> I got a chance to go back over some things that I was wrong about. (S11)

Whether or not the preservice teachers were able to transfer these experiences in perseverance to doing mathematics in non-assessment situations is not entirely clear.

The following journal responses were probably the most affirming regarding the assessments and the course as a whole:

> I plan to use [the assessments]. (S13)

> This form of assessment has also made me enjoy math more. (S2)

> Tests and quizzes aren't always the only way to find out if a student has learned something. (S5)

These future teachers experienced, most for the first time, a new approach to mathematics assessment. Because assessment was a positive experience, a majority of these preservice teachers indicated that they were likely to try alternative assessments with their students.

As expected, not all of the preservice teachers were convinced that these particular assessments represented reasonable methods for their future classrooms. Some suggested that they might use simpler versions or some other authentic task that did not reflect the work of teachers. One confessed, however, "I'm not sure if I would do this kind of assessment for my future students... the reality is the education today is way to [sic] traditional" (S12). The hope is that courses such as *Probability and Statistics*, which attempt to reflect the true mathematics through curriculum, instruction, and assessments, can change this reality.

Conclusion

The use of traditional paper-and-pencil tests to determine students' knowledge of the material presented in class is based on a behaviorist theory of learning and a view of knowledge as fixed (Romberg, 1992; Wilson, 1992, 1994a). As a result, students believe that when they get the correct answers then they have learned a great deal of mathematics (Izard, 1993). However, all they can truly claim is that their answers matched the key. Yet, paper-and-pencil tests are most familiar to preservice teachers as a method to assess mathematics achievement.

The preservice teachers taking *Probability and Statistics* took an experience typical in mathematics classrooms, test taking, and explored it in an authentic way. They did what teachers do – create tests and grade student work. In the process, these future teachers encountered mathematics assessments that required perseverance and creativity. Certainly, preservice teachers' counterproductive beliefs are well established and more than a single course will be needed to help them develop more productive beliefs (Coffey, 2000). Still, I am encouraged that these future teachers now recognize that mathematics does not have to be as they experienced it in the traditional classroom. Not only have instructional methods and materials changed, but so have the assessments. These preservice teachers encountered a different way of thinking about what it means to do and assess mathematics – a way that they can share with their future students.

References

Ball, D. L. (1990). The mathematical understandings that prospective teachers bring to teacher education. *The Elementary School Journal, 90*(4), 449-466.

Bell, K. N. (1995). How assessment impacts attitudes toward mathematics held by prospective elementary teachers. (Doctoral dissertation, Boston University, 1995). *Dissertation Abstracts International, 56,* 09-A.

Brown, C. A., & Borko, H. (1992). Becoming a mathematics teacher. In D. A. Grouws (Ed.), *Handbook of research on mathematics teaching and learning* (pp. 209-239). New York: Macmillan.

Cirulis, A. E. (1991). Three prospective elementary teachers' beliefs and the impact of their mathematics courses. (Doctoral dissertation, University of Illinois at Chicago, 1991). *Dissertation Abstracts International, 52,* 12-A.

Clarke, D. C., Clarke, D. M., & Lovitt, C. J. (1990). Changes in mathematics teaching call for assessment alternatives. In T. J. Cooney (Ed.), *Teaching and learning mathematics in the 1990s* (pp. 118-129). Reston, VA: National Council of Teachers of Mathematics.

Cobb, P., Wood, T., & Yackel, E. (1990). Classrooms as learning environments for teachers and researchers. *Journal for Research in Mathematics Education. (Constructivist views on the teaching and learning of mathematics). Monograph 4* (pp. 107-122). Reston, VA: National Council of Teachers of Mathematics.

Coffey, D. C. (2000). An investigation into relationships between alternative assessment and pre-service elementary teachers' beliefs about mathematics. (Doctoral dissertation, Western Michigan University, 2000). *Dissertation Abstracts International, 61,* 05-A.

Dossey, J. A. (1992). The nature of mathematics. In D. A. Grouws (Ed.), *Handbook of research on mathematics teaching and learning* (pp. 39-48). New York: Macmillan.

Galbraith, P. (1993). Paradigms, problems and assessment: Some ideological implications. In M. Niss (Ed.), *Investigations into assessment in mathematics education* (pp. 73-86). Boston: Kluwer Academic Publishers.

Hancock, L., & Kilpatrick, J. (1993). Effects of mandated testing on instruction. *Measuring what counts* (149-174). Washington, DC: National Academy Press.

Izard, J. (1993). Challenges to the improvement of assessment practice. In M. Niss (Ed.), *Investigations into assessment in mathematics education* (pp. 185-194). Boston: Kluwer Academic Publishers.

Lambdin, D. V., Kehle, P. E., & Preston, R. V. (Eds.). (1996). *Emphasis on assessment: Readings from NCTM's school-based journals*. Reston, VA: National Council of Teachers of Mathematics.

Lester, F. K., Jr., & Kroll, D. L. (1991). Implementing the standards: Evaluation - a new vision. *Mathematics Teacher, 84*(4), 276-284.

McLeod, D. B. (1992). Research on affect in mathematics education: A reconceptualization. In D. A. Grouws (Ed.), *Handbook of research on mathematics teaching and learning* (pp. 575-596). New York: Macmillan.

Middleton, J. A. (1999). Curricular influences on the motivational beliefs and practice of two middle school mathematics teachers: A follow-up study. *Journal for Research in Mathematics Education, 30*(3), 349-358.

National Council of Teachers of Mathematics. (1989). *Curriculum and evaluation standards for school mathematics*. Reston, VA: Author.

National Council of Teachers of Mathematics. (2000). *Principles and standards for school mathematics*. Reston, VA: Author.

Pajares, M. F. (1992). Teachers' beliefs and educational research: Cleaning up a messy construct. *Review of Educational Research, 62*(3), 307-332.

Raymond, A. M., & Santos, V. (1995). Preservice elementary teachers and self-reflection: How innovation in mathematics teacher preparation challenges mathematical beliefs. *Journal of Teacher Education, 46*(1), 58-70.

Romberg, T. A. (1992). Assessing mathematics competence and achievement. In H. Berlak, F. M. Newmann, E. Adams, D. A. Archbald, T. Burgess, J. Raven, & T. A. Romberg (Eds.), *Toward a new science of educational testing and assessment* (pp. 23-52). Albany, NY: State University of New York Press.

Schoenfeld, A. H. (1988). When good teaching leads to bad results: The disaster of "well taught" mathematics courses. *Educational Psychologist, 23*(2), 145-166.

Schoenfeld, A. H. (1992). Learning to think mathematically: Problem solving, metacognition, and sense making in mathematics. In D. A. Grouws (Ed.), *Handbook of research on mathematics teaching and learning* (pp. 334-370). New York: Macmillan.

Thompson, A. (1992). Teachers' beliefs and conceptions: A synthesis of the research. In D. A. Grouws (Ed.), *Handbook of research on mathematics teaching and learning* (pp. 127-146). New York: Macmillan.

Wheeler, D. (1993). Epistemological issues and challenges to assessment: What is mathematical knowledge? In M. Niss (Ed.), *Investigations into assessment in mathematics education* (pp. 87-95). Boston: Kluwer Academic Publishers.

Wilson, L. D. (1994a). *Assessment reforms in search of a theory.* Paper presented at the annual meeting of the American Educational Research Association, New Orleans, LA (ERIC Document Reproduction Service No. ED 372 103).

Wilson, L. D. (1994b). What gets graded is what gets valued. *Mathematics Teacher, 87*(6), 412–414.

Wilson, M. (1992). Measuring levels of mathematical understanding. In T. A. Romberg (Ed.), *Mathematics assessment and evaluation: Imperatives for mathematics educators* (pp. 213-241). New York: State University of New York Press.

David Coffey, Assistant Professor of Mathematics at Grand Valley State University, Allendale, Michigan, teaches both content and methods classes for preservice elementary and secondary mathematics teachers. His professional interests include assessment, mathematical beliefs, learning theory, and the study of probability and statistics. In order to remain grounded in the reality of public schools, he occasionally teaches with his wife, Kathy, a first-grade public school teacher. [coffeyd@gvsu.edu]

Appendix A
Assessment Writing

OBJECTIVE: to assess your ability to construct and draw inferences from plots, tables, and graphs that present data from real-world situations, and to determine the depth of your understanding

METHOD: Pairs of students will design assessment questions that could be used to evaluate their peers' understanding of the topics presented in class. The assessments should adequately represent the main concepts and procedural information found in chapters one and two of the [*Probability and Statistics*] course pack and additional information discussed during class. The assessment questions turned into the instructor will also need a correction key. It is assumed that all students "taking" your exam will have access to a TI-73 calculator. Assigning point values is not required or expected. It is expected that your group's work will demonstrate your understanding and mastery of the topics.

NEEDED: Read the following criteria several times. The final product **must** include the following in order for the project to be acceptable:

- unique questions that attend to main concepts and procedural information (one question can and should attend to several rubric criteria).

- a key to the assessment that shows possible correct solutions for each question.

- a rubric of the assessment task thoughtfully completed by the pair.

- an overall evaluation of your small group (including yourself) using the rubric provided.

GRADE: The instructor will use the attached rubric to evaluate the
                    projects. A scale for each criteria is presented below.

5 (*exemplary*) Questions AND responses clearly demonstrate deep understanding of the material and of the
intent of the assessment. The developed questions and
responses are clear in their purpose, are of a high quality,
and provide sufficient intellectual challenge to the test-
taker. The task highlights the group's creativity.

4 (*above average*) Many of the qualities from the Exemplary
category are evident but not all. There must be evidence
of a good understanding of the material and intent of the
assessment.

3 (*minimally met*) Some of the qualities from the Exemplary
category are evident.

2 (*needs work*) Unacceptable level of understanding is
demonstrated. (i.e., all work is not shown)

1 (*attempted*) Project reveals a lack of understanding of the
task and/or the material.

0 (*not evident*) No evidence provided.

Each group member should assess all work before it is
submitted and complete the following rubric.

Assessment Writing Rubric

"Instruction guide" for using the rubric:

- **READ** the rubric through carefully.
- **DEVELOP** and **ASK** questions about the scoring and expectations of using the rubric.
- **WORK** on the task, **KEEPING** the rubric close at hand.
- Critically **EVALUATE** your work using the rubric.
- **COMPARE** your evaluation of your work using the rubric with your peers.
- **REWRITE** your work based upon the critical evaluations.
- **SUBMIT** the rubric below, with your group's brief but thoughtful comments on how each expectation was met, at what level you met the expectation, and why you believe the work is at that level.

NCTM Standards

___ Questions ask students to apply knowledge and truly use problem solving skills, not simply recite answers.

___ Questions make use of "real-world" data the [*Probability and Statistics*] students would find interesting and relevant to their future teaching profession.

___ Questions provide opportunities for communicating student understanding in a variety of ways vs. simply a computational or single sentence response.

Constructing Graphs

___ Graphs correctly represent the information being presented.

___ A circle graph has been chosen for the appropriate data and I can clearly determine that it is correctly based on raw data provided.

___ A histogram has been chosen for the appropriate data and I can clearly determine that it is correctly based on raw data provided.

Constructing Tables

___ A contingency table has been chosen for the appropriate data and I can clearly determine that it is correctly based on raw data provided.

___ A frequency table has been chosen for the appropriate data and I can clearly determine that it is correctly based on raw data provided.

Constructing Plots

___ A stem-and-leaf plot has been chosen for the appropriate data and I can clearly determine that it is correctly based on raw data provided.

Interpreting Graphs

___ A good distribution of read, derive, and interpret questions are provided (this does not mean that every question must have a read, derive, and interpret component nor that other "types" of questions should not be included!). These need to be labeled on the key.

___ Questions involving a misleading (vs. just inappropriate) graph are provided.

___ Questions involving a picture graph are provided.

___ Questions involving a bar graph are provided.

___ Questions involving a circle graph are provided.

___ Questions involving a line graph are provided.

___ Questions involving a scatter plot are provided.

___ Questions involving a histogram are provided.

Interpreting Tables

___ Questions involving a contingency table are provided.

Interpreting Plots

___ Questions involving a stem-and-leaf plot are provided.

Technology Use

___ At least two questions are developed that make appropriate use of the TI-73 and its graphical/tabular features.

Overall Quality

___ Assessment questions are word processed, clearly written, and error free (the key may be hand written but needs to be legible).

___ Assessment task is submitted on time.

___ Assessment rubric has been thoughtfully responded to and submitted with the task (worth double points).

Evaluating Responses

OBJECTIVE:
- to make and evaluate arguments that are based on data interpretation;
- to understand and apply concepts of central tendency, spread, and variability;
- to describe, in general terms, the normal curve and its properties to answer questions about sets of data that are assumed to be normally distributed;
- to construct and draw inferences from tables and graphs that represent data.

METHOD:
In pairs, students will create a key for a pre-existing [*Probability and Statistics*] exam. The exam questions are based on material presented in chapters two through four from the course pack, and supplemental materials presented in class.

Individually, students will use their "pair" key to evaluate the answers from an exam completed by a student who was selected randomly from a large pool of imaginary students.

This last piece of the assessment will be completed in class and will involve the whole class period.

NEEDED:
The final product **must** include the following in order for this assessment to be acceptable:
- a key for the original exam from each pair (one key is needed from each pair; during the assessment, everyone will need an individual copy of the key, making a total of 3 copies);
- a copy of the evaluated exam from each individual in the pair.

GRADE:
I will use the attached rubric to evaluate the assessment. You should self-assess your key and each pair should assess their evaluation of the answers before it is turned in to the instructor.

Evaluating Responses Rubric

Key (*0-40 points*)
Point values have been assigned to each problem and are provided on the exam itself next to each question. I will be "grading" your key using the following criteria as I evaluate each response:

100%   **Exemplary** – all work is shown and correct. Construction of plots is correct. Answers requiring an explanation are clear, concise, and correct.

80%   **Above average** – all work is shown and the student is headed in the right direction, but some steps are incorrect. Construction of plots contains an error. Explanation is not always clear and concise.

60%   **Minimal** – all work is shown but method chosen to solve the problem is not appropriate, or some work is missing. Construction of plots contains two errors. An explanation has some elements of understanding but result is unclear.

40%   **Needs work** – no work is shown, but answer is correct. Construction of plots contains three errors. The explanation has one point of understanding but the result is incorrect.

20%   **Attempted** – the problem has been tried, but contains few elements of understanding.

0%   **Blank or Not Attempted** – the problem has not even been attempted or the problem has been tried, but contains no elements of understanding.

Evaluation and Summary of the Imaginary Student Work (*0-60 points*)
In class, each of you will use the key you created to evaluate a [*Probability and Statistics*] student's exam. You are to use the above scale as you evaluate the work of the student. I will then assess your evaluation work based on the following:

> The point evaluation of each response is reasonable based on the number of errors present. In other words, all mistakes have been identified. (*0-25 points*)

> If less than full credit is given to an answer, then a justification (using the rubric) for the score is provided for the student on the exam itself. (*0-20 points*)

> Possible rationales for student mistakes are provided for the appropriate questions. (*0-15 points*)

Mathews, S. M.
AMTE Monograph 1
The Work of Mathematics Teacher Educators
©2004, pp. 67-85

5

The Experiences in a Concepts in Calculus Course for Middle School Mathematics Teachers

Susann M. Mathews
Wright State University

Both the Mathematical Association of America and the Conference Board of the Mathematical Sciences have recommended that middle-grades teachers have experiences with the underlying themes of calculus. This paper describes a course in the concepts of calculus that provides such experiences. I focus on the experiments I have created to help teachers construct an understanding of the big ideas of calculus: limits, derivatives, and integrals. I include the types of questions I ask to help teachers construct meaning for these ideas, including details of writing assignments and oral interviews.

Concepts in Calculus for Middle School Teachers is a mathematics course for preservice teachers at Wright State University (WSU). *A Call for Change* from the Mathematical Association of America (MAA) (Leitzel, 1991) recommended five areas to be included in the mathematical preparation of middle-grades mathematics teachers: number concepts and relationships, geometry, algebra and algebraic systems, probability and statistics, and concepts in calculus. Thus, when we at Wright State University began developing a cohesive program for preservice middle-school mathematics teachers, we created six courses to address these recommendations. In this paper, I discuss the course I developed to provide our preservice teachers with experiences in the concepts in calculus. *A Call for Change* (Leitzel, 1991) states that preservice middle-grades mathematics teachers should:

- interpret, with the aid of graphs, diagrams, and physical models, the concepts of limit, differentiation, and integration, and the relationships among them;

- construct concrete examples of finite sequences, extend the ideas to infinite sequences and series, relating them to the meaning of approximation of nonterminating decimals and to the approximation of functions;
- explore concrete realistic problems involving average and instantaneous rates of change, areas, volumes, and curve lengths, and relate these problems to concepts of differentiation and integration. (p. 24)

In the *Concepts in Calculus* course, the focus is on recommendations 1 and 3, because ideas relating to recommendation 2 are included in the course *Mathematical Modeling for Middle School Teachers*.

The Mathematical Education of Teachers (Conference Board of the Mathematical Sciences [CBMS], 2001) does not have a specific section on concepts in calculus for teachers of grades 5 – 9; however, underlying themes of calculus occur in the sections Numbers and Operations, Measurement and Geometry, and Algebra and Functions. For example, students need to be able to "make sense of large numbers and small" (p. 28), "decompose and recompose non-regular shapes to find areas or volume" (p. 111), and "develop a deep understanding of variables and functions" (p. 108), including a study of the mathematics of change. Preservice teachers frequently have no idea why they should take such a course when they are preparing to teach middle school. Therefore, as we perform and analyze the experiments described in this paper, I explicitly tie differentiation to rates of change and integration to finding the area between a function and the x-axis. I explain that their students will be studying rates of change when they study the slope of linear functions and will be preparing for calculus when they find the area or volume of non-regular shapes. As we work with limits, repeatedly creating smaller intervals for approximating integration and differentiation, we discuss how their students could begin working with this idea. Middle-grades students could approximate the area of a non-regular surface by using course grid paper; they could then make better approximations by using finer and finer grid paper, comparing the results of each approximation.

As the course progresses, I ask the teachers how the course content ties to the standards for grades 6 – 8 of the *Principles and Standards for School Mathematics* (*PSSM*) (National Council of Teachers of Mathematics [NCTM], 2000). After taking this class, one in-service middle-grades teacher wrote, "The most important thing I learned from this class is that what I do with my 7th and 8th graders will come back to help them in advanced mathematics. It just struck me in a fresh way how all of math is related … very cool."

Concepts in Calculus has two main conceptual themes: (a) the motivation and need for calculus and (b) the concepts of limit, differentiation, and integration and the relationships among them (Leitzel, 1991). A problem-solving approach is used to address both themes, focusing on making sense of the mathematics and its connections rather than working on the symbol manipulation skills of differentiation and of integration.

Preservice secondary mathematics teachers at WSU are required to take two of six mathematics courses for middle-school teachers. The big ideas of the *Concepts in Calculus* course are fundamental to the understanding of calculus, yet many individuals who succeed in traditional calculus courses still profess not to understand them (Dudley, 1993). Therefore, I highly recommend to the future secondary teachers that they choose the *Concepts in Calculus* course as one of their two courses.

The use of the experiments and hands-on learning activities described in this paper help both preservice and inservice teachers construct an understanding of the big ideas of calculus. They begin with the concrete, move to the pictorial, and finally proceed to the abstract. After grading their work, my graduate assistant wrote, "For the most part though, I think most of your class had an excellent grasp of the FTC [Fundamental Theorem of Calculus], and a good idea of integral, derivative, and limit concepts, and I would put them up against any Calc III class as to these general concepts." One preservice secondary teacher addressed how the course will benefit his teaching when he wrote that he learned "concrete ways in which to express these concepts and ideas (labs, projects, etc.) other than just showing students how to do calculus symbolically."

The first half of the course is devoted to constructing the ideas of limit, derivative, and integral and their relationships. Teachers conduct three experiments, the first two of which graphically develop the Fundamental Theorem of Calculus. After the first experiment, teachers write up their work and explain their new understandings. After the second, I give them each an oral interview. In the third experiment, teachers revisit the limiting process and again write up their work. The text, *Conversational Calculus* (Cohen & Henle, 1997), is used to introduce vocabulary and notation, to reinforce the ideas developed during the experiments, and to introduce differential equations and their applications.

Experiment One: Fundamental Ideas of Calculus

On the first day, each group gets a non-cylindrical vase, a burette with a stand, a ruler, and water. (See Appendix A for a copy of the experiment.) The group predicts what a graph of volume vs. height would look like for its vase and then proceeds to fill the vase with water, keeping track of the volume at 2 mm increments of height. Collecting data and determining precisely how to record it takes most of the first class. (The class meets 1 hour and 40 minutes twice a week.) As homework before the next class, teachers make three tables, and begin making three sets of graphs, first by hand and later using technology:

1. Graphs of volume as a function of height, $V(h)$ vs. h, for every 16 mm, every 8 mm, and every 4 mm;
2. Graphs that approximate the derivative with large intervals to small using the $\dfrac{\Delta V}{\Delta h}$ column from each table;
3. Graphs that approximate the areas between derivative graphs and the x-axis.

Teachers are often confused about what to graph and time is well spent doing the graphs in class after teachers have attempted the homework. Using the graphs in Figure 1 (p. 72), we discuss the original function, the derivative and its meaning, the integral of the derivative and its meaning, and the idea of a limit as the size of the intervals is reduced. After groups have completed their graphs, I ask such questions as the following:

- Why did you make your prediction about the graph of volume vs. height?
- How do your graphs of volume vs. height compare to your prediction?
- Is volume a function of height? Why or why not? Do you know its symbolic representation?
- Must a function have a closed-form equation to be a function?
- What are the units of each axis on each set of graphs? (Many struggle with the units on the graphs of the derivative and the integral of the derivative; requiring units helps make what is happening concrete. For example, when teachers realize that the graphs approximating the derivative have units of ml/mm vs. mm, they understand that derivatives are rates of change. The fact that the graphs of the integral of the derivative have units of ml vs. mm, as in the first graph, helps teachers recognize that the result is the original graph.)
- Why do we first approximate what is happening with large intervals of height and work toward small intervals?
- What does the rate of change of the function mean?
- Why do the graphs of the approximations to the derivative get more varied as the intervals decrease but the approximations to the integral graphs get smoother?
- What does the derivative graph tell about the original function? (After determining the units for the approximation to the derivative graph, teachers reply that these graphs tell the rate of change of the growth in volume as a function of the height of the vase.)
- Why does the area under the derivative graph, that is, the integral of the derivative, look like the original function?

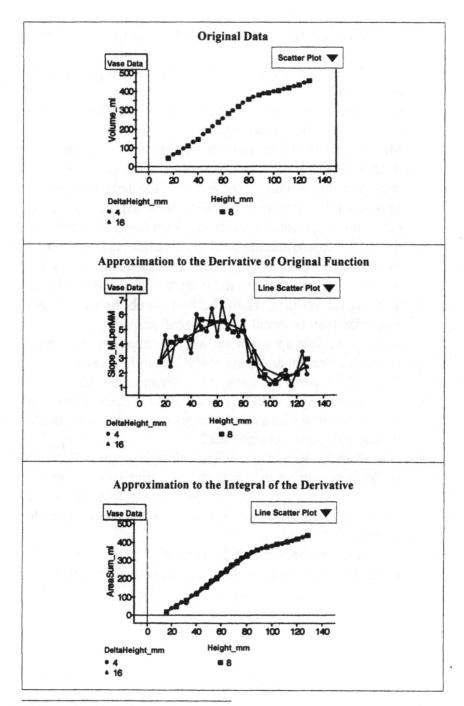

Figure 1. Graphs for Experiment 1

As teachers try to understand what the units will be, we discuss how they determined the function values for the area under the derivative graph: they multiplied ml/mm × mm to get ml. By the time teachers write about this experiment, most have realized that they have multiplied the rate of change (ml/mm) by how long the function has been changing (in terms of change in height in mm). Whole-class discussions follow the group discussions, leading to the verbalization of the Fundamental Theorem of Calculus.

Many preservice secondary mathematics teachers who have had as many as four quarters of calculus are surprised when they see how the integral of the derivative results in the original function plus a constant of zero. They often exclaim that they finally understand the Fundamental Theorem of Calculus, despite previously working problems symbolically using the theorem. They also understand the ties to geometric ideas by seeing that the slopes of chords, $\frac{\Delta V}{\Delta h}$, approach the slope of the tangent, $\frac{dV}{dh}$, as Δh is decreased. For the preservice middle-grades mathematics teachers, this experiment provides an immediate introduction to the big picture of functions, derivatives, integrals, and limits and their importance. Throughout the first part of the course, we concentrate on the graphical and tabular representations of the approximations. Only later in the course do the preservice teachers work on calculating symbolically.

After this experiment and subsequent discussions, the teachers write a report that includes a clear statement of the reasons for the experiment and its goals, a summary of their work, tables of data, graphs, a comparison of their results to their prediction, and a summary of the results with explanations of new understandings and the relationships discovered. The act of writing precipitates continued discussions as teachers struggle to communicate their conceptualization of the relationships among a function, its derivative, and the integral of the derivative.

Experiment Two: Connecting a Function,
Its Derivative, and Its Integral

Experiment Two, which preservice teachers complete individually, reinforces the big ideas of the first experiment as teachers (1) graph a piece-wise defined linear function, (2) graph the area between the original function's curve and the x-axis, (3) graph the slope of the original function, and (4) graph the area between the derivative and the x-axis. These graphs are shown in Figure 2 (p. 75); the complete experiment is in Appendix B.

The first time I gave this work, I simply asked the preservice teachers to write a report as they had after the first experiment. However, teachers had difficulty interpreting the graphs and expressed concern about what they were supposed to write. Thus, over time I developed the following questions that I ask teachers to discuss in their groups and then discuss with the whole class.

1. What kind of graph is the original function?
2. What did you find when you found the area between that function and the x-axis? What kind of function does it resemble?
3. Why is the slope graph of the original function two straight lines? Why are there two lines instead of one? What do they tell about the original function? What is another word for the slope of this function?
4. Does the final graph of the area between the derivative and the x-axis resemble any of the other graphs? If it does, which one and why? If it doesn't, why not? Why isn't it exactly like the original graph?

For those who have never had calculus, these graphs suggest that the integral of a linear function is quadratic, the integral of a constant function is linear, and the derivative of a linear function is a constant. For those who have previously had calculus either in high school or as secondary mathematics majors, this experiment often leads to "Oh, so that is what is going on!" One middle-grades preservice teacher, whose writing is representative of her peers, wrote, "Start with a function and find the integral of the derivative and it goes back to the function plus a constant. This follows the fundamental theorem of calculus." Writing is part of helping these preservice teachers clarify their ideas before I give individual interviews as an assessment, using the questions in Appendix C.

Each of the 30–35 preservice teachers schedules a 20-minute interview with me. I have found that the individual interviews have been an opportunity for both the preservice teachers and me to determine what concepts they understand and what concepts still pose difficulties. As suggested by Clarke (1997), these interviews offer insights into perceptions, conceptualizations, and understandings. I grade each answer based upon its mathematical correctness and each explanation upon its accuracy, detail, and clarity. When assigning a grade, I ask the individual to evaluate his or her own performance and understanding, using the following general criteria:

A – The preservice teacher has solid control of the material, not just by memorization but with the ability to think, and use techniques to solve problems.

B – The preservice teacher is usually in control of the material and has good knowledge.

C – Although the preservice teacher can complete the work most of the time, he or she is sometimes confused.

D – The preservice teacher has some knowledge of the material but is often confused or unsure of himself or herself.

F – The preservice teacher lacks knowledge and demonstrates little or no control of the material.

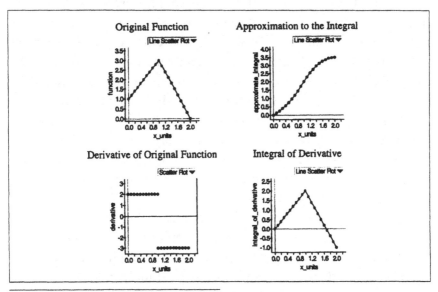

Figure 2. Graphs for Experiment 2

As future teachers, it is imperative that they understand what they know, how well they understand it, and what they do not know. Although almost all of the preservice teachers expressed dismay at the requirement to evaluate themselves, almost all of them have been within a plus or minus of the grade I had decided to assign.

Experiment Three: Maximum Volume of a Cone

In the third experiment, preservice teachers revisit the limiting process in approximating an integral, connect geometry with algebra and calculus, and explore a function, its maximum, and its derivative. I hand each group scissors and a set of paper disks with equal radii and ask them to determine the angle that results in a cone with the largest volume when a sector is removed from the disks. As in the first experiment, I ask the teachers to predict the angle that they should remove to create the largest volume and to explain their prediction. They cut out a sector of each disk and wrap the remaining sector into a cone. The directions for this experiment are provided in Appendix D.

I direct them to measure the volume of each cone with 2 cm × 2 cm × 2 cm blocks, with 1 cm × 1 cm × 1 cm blocks, and with rice, followed by making a table of their results. Most groups begin to cut out angles from 30° to 180°, systematically creating 10 cones. However, some groups cut out three to five seemingly random angles, make their cones, and then state the angle remaining in their largest cone. At that point, I ask them what happens to the volume of the cones with angles whose measures are between those of the angles they used. When they determine that they do not know, they systematically make more cones. After graphing their results on the same axes, they discover that their results vary with different intervals of measure. We relate their results to the idea of getting better approximations to a derivative and to an integral by using smaller-width intervals, with a limiting width of zero giving the most accurate result. Working with teachers who have been through our program, we used this experiment as a basis for a unit on volume and area for middle-school students. Thus, the preservice teachers see that the underpinnings of calculus can begin in the middle grades.

We then discuss the rate of change of volume with respect to the angle remaining, $\dfrac{\Delta V}{\Delta \theta}$, and its relation to derivatives. At this point I ask the teachers to derive the formula for the volume of their cones in terms of the angle θ that remains. They know that the formula for the volume of a cone is $\dfrac{1}{3}\pi r^2 h$, where r is the radius of the base of the cone and h is the height of the cone. However, because they do not know r and h, I guide them to solve for both r and h in terms of the slant height, which is the radius of the original circle, and θ, which is the angle remaining. Thus, the teachers work with function notation, relating the change of one variable as the result of the change of another. The final formula as a function of θ in radians and the cones' slant height R (the radius of the original disk) is

$$V(\theta) = \frac{\pi R^3}{3}\left(\frac{\theta}{2\pi}\right)^2 \sqrt{1 - \left(\frac{\theta}{2\pi}\right)^2}$$. Figure 3 (p. 78) contains a graph of the volume vs. the remaining angle measurement (in degrees) using all three measurement units (large blocks, small blocks, and rice) as well as a graph of the volume function in closed form.

To graph their work in *Mathematica*, teachers must convert their formula from measuring θ in degrees to measuring θ in radians. Although they have been introduced to radians in their mathematics for elementary teachers courses and in their geometry course, most have only a tenuous hold on the concept. The task of graphing provides another opportunity to revisit the concept to help teachers construct a solid understanding. Once teachers have graphed their functions using *Mathematica* and compared them to the graphs based on their data, I ask them to graph the derivative and explain its meaning as the angle moves from 0 radians to 2π radians along the x-axis. There is generally an "aha!" after realizing the value of the derivative is zero when the function takes its maximum value, even in real-world problems. After concluding the experiment and class discussion, teachers again write a report about the experiment.

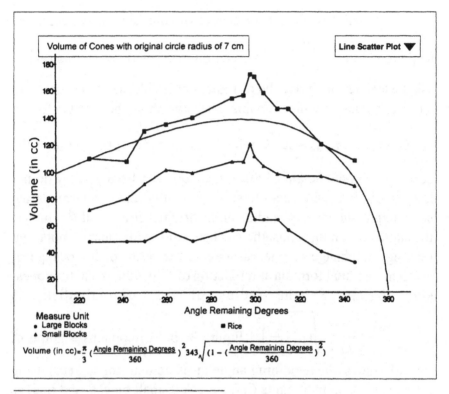

Figure 3. Graphs for Experiment Three

Other Course Activities

Between experiments, we work from *Conversational Calculus*, whose premise is that calculus is a language used to describe the universe. This text provides practice and reinforcement of the vocabulary and notation, all of which are new to teachers who have not studied calculus before. Teachers use *Mathematica* to investigate graphical representations of functions, derivatives, and integrals and to solve simple applied problems. Expanding their use of *Mathematica* to include its computer algebra system, teachers solve elementary differential equations, explore the graphs of the differential equations and their solutions, and investigate a few of their applications. To complete the first half of the course, they solve "A Murder of Quantity" (Cohen & Henle, 1997, pp. 194-195), in which they solve a murder mystery by applying Newton's Law of Cooling to the body of the victim.

Conclusions

In the contexts of filling vases with water; exploring linear graphs, their derivatives, and their integrals; and investigating cones, teachers work on making sense of the concepts of limit, differentiation, and integration, and the relationships among them. Additionally, they solve real-world problems involving differential equations.

As I developed *Concepts in Calculus for Middle School Teachers*, I decided that the course needed to address the three big ideas of calculus – limit, differentiation, and integration – and their applications. I searched through several calculus texts before finding one that concentrated on these concepts without quickly turning to symbol manipulation. After finding *Conversational Calculus*, which supplied the verbal and graphical approach, I still needed to develop experiences for my preservice teachers that would make these ideas concrete. As the course evolved, I appreciated yet again that it is much more difficult to provide worthwhile problems and pose meaningful questions than it is to explain concepts well.

Hints for Starting Up

If other teacher educators want to begin a course in *Concepts in Calculus*, realize that your preservice or inservice teachers are constructing understanding inductively rather than deductively; it takes several experiments and writing for them to recognize and value the patterns and relationships. Furthermore, they need to study the concepts from as many viewpoints as possible, such as working from derivatives to integrals and back to derivatives or beginning with integrals and working from there to derivatives and back to integrals. As a practical note, gather the equipment and completely work through the experiments to discover some of the things that may go wrong. For example, if the burettes are not firmly mounted to the ring stands, they may tip over and break or paper towels may be needed because water gets spilled. Most chemistry departments have burettes with chipped tips that are too damaged for accurate chemical analyses, but that are adequate for the first experiment. Make arrangements with the laboratory technicians well in advance so that they will save these burettes for you instead of throwing them away. Also remember that everything takes longer than anticipated if it is to be meaningful and successful. Instructors must not get impatient with class members as they strive to figure out what to do and what the relationships are. It is easy to help them too much. However, it is through the struggle that they really learn.

References

Clarke, D. (1997). *Constructive assessment in mathematics.* Berkeley, CA: Key Curriculum Press.

Cohen, D. W., & Henle, J. M. (1997). *Conversational calculus, preliminary edition* (Vol. 1). Reading, MA: Addison-Wesley.

Conference Board of the Mathematical Sciences. (2001). *The mathematical education of teachers, part II.* Providence, RI: American Mathematical Society.

Dudley, U. (Ed.). (1993). *Readings for calculus.* 5 vols. Vol. 5, *Resources for Calculus.* Washington, DC: Mathematical Association of America.

Leitzel, J. (Ed.). (1991). *A call for change: Recommendations for the mathematical preparation of teachers of mathematics.* Washington, DC: Mathematical Association of America.

National Council of Teachers of Mathematics. (2000). *Principles and standards for school mathematics.* Reston, VA: Author.

Susann Mathews, Associate Professor at Wright State University, taught both junior high and high-school mathematics before becoming a mathematics teacher educator in 1987. Her interests lie in mathematical modeling from ancient times through the present, including how modeling can be used to make connections with other disciplines and with the world. At Wright State University, she has been a part of a team that has created a reform-based mathematics program for middle-grades teachers. She is now working with other mathematicians in the department to create such a program for secondary teachers. Her research is focused upon how Wright State University's middle-grades preservice teachers grow mathematically. [susmathews@woh.rr.com]

Appendix A
Handout for Experiment One –
The Fundamental Ideas of Calculus

Materials: For each group of four students: a non-cylindrical vase, a burette with a stand, a ruler that fits into the vase, and water.

1. Fill your vase with water from the burette, recording the volume for every 2 mm change in height, beginning at 8 mm.

2. Make three tables from your data, following the format for the first table (see page 82) for all three tables.
   a. Make a second table using $h - 4$ and $h + 4$ with $\Delta h = 8$, beginning at $h = 12$.
   b. Make a third table using $h - 2$ and $h + 2$ with $\Delta h = 4$, beginning at $h = 10$.

3. Plot V vs. h using data from each table, making 3 individual plots, keeping the scale the same for each graph. Connect the points on each graph.

4. Now plot $\dfrac{\Delta V}{\Delta h}$ vs. h for each table. You will have three graphs of slopes; keep the scales the same for all three graphs.

5. Approximate the area under the curve in each of your plots of slopes graphically. You will need to approximate the area between each two consecutive sets of data points. Record these values and graph their accumulating areas vs. h, using the same scales you used in step 3.

6. How would the initial three graphs look if you had kept constant the amount of volume you added to the vase each time and measured the height, plotting h vs. V?

Table for Recording Data from Experiment One

h [mm]	V [ml]	h - 8	V(h - 8)	h + 8	V(h + 8)	Δh (h+8) - (h-8)	ΔV V(h+8) - V(h-8)	$\frac{\Delta V}{\Delta h}$
16								
32								
48								
64								
80								
96								
112								
128								
144								
160								

Appendix B
Handout for Experiment Two –
How Are a Function, Its Derivative, and Its Integral Related?

1. In the middle of your graph paper, graph the function: $f(x) = 2x + 1$ for $0 \leq x \leq 1$ and $f(x) = -3x + 6$ for $1 \leq x \leq 2$. Let each square on your graph paper represent 0.1 on the x-axis.

2. Make a vertical line on your graph from the x-axis to the function at each 0.1 from 0 to 2. Calculate the area of each trapezoid, formed by the x-axis for each 0.1 unit, the function over the same step, and the vertical lines from the x-axis to the function. Use this information to make a table with three columns: x_i; Area from x_{i-1} to x_i; Total area from 0 to x_i.

   Make a graph at the top of your graph paper with the same scale for the x-axis as the original graph and the entries from the last column of your table for the y values.

3. Make a graph on the bottom of your paper with the same scale for the x-axis as before, but use the slope of the original function at each x_i as the y values.

4. Make a graph of the accumulated area between the x-axis and the graph from part 3 at each x_i, using the same scale as you did for the original graph.

Appendix C
Questions for Interviews about Functions, Derivatives, and Integrals

1. What type of function have you graphed? What representations have you used?

2. The original function:
    a. What is the function's rate of change?
    b. How do you know?
    c. What representations have been used? What do they tell you about the function?

3. What does the integral of the original function represent?

4. What would happen if you differentiated that integral?

5. What happened when you *integrated* the derivative of the original function? Why?

Appendix D
Handout for Experiment Three –
The Maximum Volume of a Cone

1. Given a circle with a specified radius (in this case about 7.5 cm). Predict the size of the central angle that produces a cone with the largest volume when a sector is removed from the disks. Explain why you chose the angle you did and state the size of its corresponding remaining angle.

2. Now find which central angle remaining produces a cone with the greatest volume.

3. Measure the volume of each cone with 2 cm × 2 cm × 2 cm blocks, with 1 cm × 1 cm × 1 cm blocks, and with rice. Graph those results on the same axes. What units are on each axis? How do these results compare – how are they the same and how are they different? If they are different, why are they different?

4. Discuss the rate of change of volume with respect to the angle remaining, $\dfrac{\Delta V}{\Delta \theta}$.

5. Derive the formula for the volume of a cone with the central angle remaining as the independent variable. Graph that function using *Mathematica*.

6. Discuss how this graph compares to the graphs based on your measurements. What remaining angle (after a sector is removed) gives the cone with the greatest volume? How does this compare with your prediction? With your measurements?

7. Explain how the derivative (i.e., its rate of change) of the closed-form of the function with respect to the angle remaining, $\dfrac{dV}{d\theta}$, compares with the rate of change discussed in question 4, $\dfrac{\Delta V}{\Delta \theta}$.

8. Using *Mathematica*, graph the derivative of your analytic function on the same axes as the function. Compare the graph of the derivative with the derivative as a rate of change from (7). What happens to the derivative when the function reaches its maximum value?

Contreras, J. N. and Martinez-Cruz, A. M.
AMTE Monograph 1
The Work of Mathematics Teacher Educators
©2004, pp. 87-101

6

Drag, Drag, Drag: The Impact of Dragging on the Formulation of Conjectures within Interactive Geometry Environments[1]

José N. Contreras
University of Southern Mississippi

Armando M. Martínez-Cruz
California State University, Fullerton

In this paper we describe experiences in which prospective and inservice elementary and secondary mathematics teachers have used Interactive Geometry Environments to investigate complex conjectures. One challenge with such investigations is the need to drag flexible points systematically to find and test more than one critical case of the potential solution. Otherwise, teachers have formulated incomplete or incorrect conjectures.

Learning to formulate conjectures is a complex process that typically involves stating an initial conjecture and then revising or refining it. As teacher educators, we need to provide experiences in which teachers can learn to formulate conjectures. This is important for two reasons. First, conjecturing is a critical component of doing mathematics at all educational levels, including teacher education (National Council of Teachers of Mathematics [NCTM], 2000; Polya, 1954). Second, teachers need to know how to formulate conjectures

[1] This article was partially supported by internal funds granted to the first author from an institutional grant (Grant No. P342A000029) through the U.S. Department of Education's "Preparing Tomorrow's Teachers to Use Technology" program. The opinions expressed here are those of the authors and not necessarily those of the federal government.

in order to design learning environments in which their students formulate conjectures as part of doing mathematics.

Geometry provides opportunities to engage students in formulating conjectures, facilitated by *Interactive Geometry Environments* (IGEs), such as *Cabri Geometry II* (Texas Instruments, 1998) and *The Geometer's Sketchpad®* (Jackiw, 2001). IGEs free the user from performing time-consuming activities such as computations, constructions, and repetitive tasks using paper-and-pencil procedures. More strikingly, they enable the user to manipulate or move (i.e., drag) elements of geometric configurations dynamically while preserving invariant mathematical relationships. However, we have noticed that both preservice and inservice mathematics teachers do not always take full advantage of the construction and dragging capabilities of IGEs, resulting in formulation of conjectures that often fail to include needed restrictions, boundaries, or all plausible instances for which they may be true.

In this paper, we discuss features of IGEs related to the conjecturing process and then use three examples to illustrate how insufficient dragging affects the formulation of conjectures within these interactive geometry environments. We also discuss implications of our experiences in conjecturing within IGEs for learners at all levels (e.g., teacher educators, teachers, and high school students). Because this paper focuses on the importance of conjecturing within IGEs, we do not provide proofs for the conjectures.

Features of IGEs Related to Conjecturing

IGEs provide learners with opportunities to discover and investigate patterns and formulate mathematical conjectures. However, as with any technological tool, IGEs must be used appropriately. Within IGEs, students need to learn about *construction* and *dragging*; in addition, *induction, generalization,* and *critical cases* are three other concepts that are central to investigating conjectures.

Construction

Together with other authors (e.g., Hoyles & Noss, 1994), we have observed that some learners do not distinguish easily between *drawing* and *constructing* when working within IGEs. As teacher educators, one of our first tasks is to help learners understand the difference between constructed figures and drawn figures. The constructed elements of a figure maintain the intended relationships when objects are dragged. In contrast, the intended properties of a drawn figure are not maintained when dragging flexible objects of the figure. For example, consider the two squares shown in Figure 1. When vertex D of square DRAW is dragged, the figure remains a quadrilateral but not always a square (Figure 2). In contrast, when vertex C of CONS is dragged, the square relationship is maintained (See Figures 1 and 2). The differences occur because DRAW was drawn using horizontal and vertical segments and estimating the length of each side of the quadrilateral; quadrilateral CONS was constructed, using a series of 90° rotations, starting with segment $\overline{CO}$.

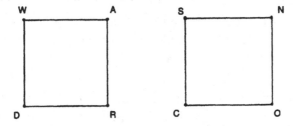

Figure 1. Two squares to illustrate drawn (DRAW) versus constructed (CONS) figures

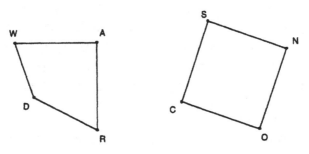

Figure 2. The result of the squares in Figure 1 after being dragged

Dragging

Another aspect of using IGEs appropriately is dragging systematically. Some learners tend to formulate a conjecture based on a single computer-generated example. As a consequence, they may formulate conjectures that do not consider all possible conditions, needed restrictions, or all plausible instances for which they may be true. In this paper, we identify such conjectures as *incomplete conjectures*.

We have noticed that learning to drag involves two cognitive tasks: learning what objects are "dragable;" and learning to drag systematically rather than randomly. With almost any examples, learners notice that some objects are "dragable" while others are "non-dragable" (Hölzl, 1996). Some IGEs (e.g., The Geometer's Sketchpad) allow the user to drag "non-dragable" objects while others (e.g., Cabri Geometry) do not. In the case of The Geometer's Sketchpad, although all objects can actually be dragged, dragging "non-dragable" objects moves all elements of the figure defined by the objects and the underlying relationships are unaffected. For example, consider square CONS constructed from segment $\overline{CO}$ (see Figure 1). Vertex C is dragable, meaning the figure changes in size or orientation. However, vertex S is non-dragable, meaning that the entire figure undergoes a translation but no other changes occur in the figure. Dragable points are considered to be *flexible* points in the discussions that follow.

To drag systematically, therefore, means to drag, in a certain order, every flexible point of a figure to determine invariant relationships that exist as a figure changes. To facilitate this learning, teachers complete several explorations in which insufficient or random dragging may lead to incomplete conjectures. Through these guided explorations in which they distinguish "dragable" and "non-dragable" objects, many teachers begin to understand the need to drag every flexible point of a figure systematically as they test their conjectures.

Induction and Generalization

Learners use an inductive process to derive a conjecture because they observe a finite number of specific empirical examples. As learners verify that the conjecture is true for all observed examples, they try to extend the conjecture to all plausible cases. This process of extending an observed relation in a finite number of examples to a larger set of examples is *generalization*. Thus, the result of a generalization is only a plausible mathematical conjecture that must then be proved. In fact, from the explorations teachers often "discover" generalizations that are well known theorems.

Critical Cases

Because the experimental evidence supporting a conjecture comes from a finite number of cases, the conjecturer needs to examine all critical cases and look for counterexamples or disconfirming evidence. A critical case is a condition or restriction on a figure or a set of points related to the figure that represents a special, extreme, plausible, or negative instance for the given exploration. A conjecturer who fails to find and test relationships for all critical cases of the geometric figure may formulate false conjectures for which the observed relationship does not hold. Learners often need to refine a conjecture to accommodate all critical cases and any exceptions.

Within IGEs, a teacher often asks students to make or use a construction to examine or discover some underlying relationships. Students normally use the measurement or construction capabilities of the IGE to support or refute such relationships. Next, students drag flexible points of the geometric figure to determine if a relationship remains invariant for a large sample of such figures. During this stage of the investigation, the conjecturers must exhaust the full range of critical cases to make a plausible generalization based on a finite sample of instances. Then, students formulate a conjecture with note of any exceptions and that includes all possible instances for which it may be true; in this paper, we refer to such a conjecture as a *complete conjecture*.

Examples to Illustrate Conjecturing Within IGEs

In this section, we use three examples to illustrate the use of the features of IGEs related to conjecturing. We sometimes start with an example based on Napoleon's configuration to help learners recognize the importance of dragging systematically to find and test all critical cases when formulating generalizations. Then we consider problems relating to equilateral triangles and to Vivani's problem.

Segments in Napoleon's Configuration

The configuration shown in Figure 3 consisting of △ABC and the equilateral triangles formed on each side (△ABD, △BCE, and △CAF) is known as Napoleon's configuration. What special properties exist for the segments $\overline{AE}$, $\overline{BF}$, and $\overline{CD}$? As suggested by the figure, $\overline{AE}$, $\overline{BF}$, and $\overline{CD}$ seem to be congruent and concurrent. However, upon dragging one of the vertices of △ABC, learners may notice that the segments still seem to be congruent but they are no longer always concurrent (Figure 4). To formulate a complete conjecture, learners need to investigate the critical cases for which the segments are still concurrent. If they slowly drag a vertex of △ABC until they obtain the critical case for which the segments are still concurrent, they may obtain a triangle such as the one displayed in Figure 5. To determine any special characteristics of △ABC, learners measure its sides and angles and find that the segments are still concurrent for a triangle having an angle measuring 120°. They can *construct* the configuration for a triangle having an angle measuring 120° at vertex B and drag flexible points (any vertex of △ABC) to confirm the conjecture; they can also reason that $m(\angle ABC) = 120°$ because the three segments are concurrent at vertex B of △ABC.

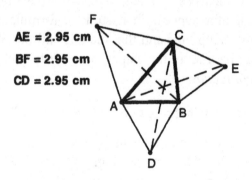

AE = 2.95 cm
BF = 2.95 cm
CD = 2.95 cm

Figure 3. $\overline{AE}$, $\overline{BF}$, and $\overline{CD}$ seem to be congruent and concurrent

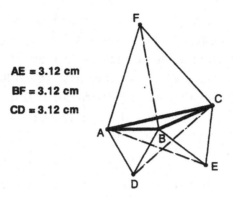

Figure 4. $\overline{AE}$, $\overline{BF}$, and $\overline{CD}$ are not always concurrent

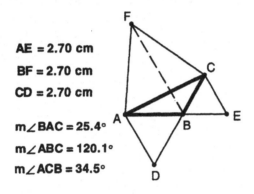

Figure 5. The critical case for which $\overline{AE}$, $\overline{BF}$, and $\overline{CD}$ are still concurrent

The final conjecture can be stated as follows:

Let $\triangle ABC$ be any triangle. Construct outward equilateral triangles, $\triangle ABD$, $\triangle BCE$, and $\triangle CAF$ on the sides of $\triangle ABC$. Segments $\overline{AE}$, $\overline{BF}$, and $\overline{CD}$ are congruent for any triangle. Furthermore, the segments are concurrent for triangles whose greatest angle measures 120° or less.

A triangle with an angle of 120° is an instance of a critical case because it is the extreme triangle for which the conjecture still holds. For any triangle having an angle measuring more than 120°, some parts of the conjecture, such as concurrency, are no longer true. At this point in the investigation, the instructors ask learners if it is possible to extend the conjecture about concurrency to any triangle. Although most learners say "no" or do not respond, others notice that the lines $\overleftrightarrow{AE}$, $\overleftrightarrow{BF}$, and $\overleftrightarrow{CD}$ are concurrent even though the segments are not (Figure 6). In this example, there are three critical cases to investigate: A triangle with all interior angles measuring less than 120°, a triangle with one interior angle of measure 120° (an extreme case), and a triangle having one interior angle measuring more than 120°. As the example illustrates, learners need to find and test the three critical cases before formulating their conjecture in order to note any restrictions for which the conjecture is plausible.

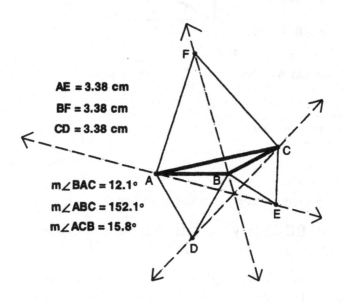

AE = 3.38 cm

BF = 3.38 cm

CD = 3.38 cm

m∠BAC = 12.1°

m∠ABC = 152.1°

m∠ACB = 15.8°

Figure 6. Concurrency of $\overleftrightarrow{AE}$, $\overleftrightarrow{BF}$, and $\overleftrightarrow{CD}$

Points on the Circumcircle of an Equilateral Triangle

In this example, we investigate whether there is a relationship among the distances between a point on the circumcircle of an equilateral triangle and the vertices of the triangle. Let P be a point on the circumcircle of equilateral $\triangle ABC$ (Figure 7). The initial conjecture may be that $PC = PA + PB$. As learners drag point P along the circumcircle, they find that $PC = PA + PB$ for several points on the circumcircle but not for others (Figure 8). At this point they may investigate the points P for which $PC = PA + PB$. As learners slowly drag point P along the circumcircle in both directions, they notice that A and B are the extreme cases for which $PC = PA + PB$. For any point "beyond" A or B, the initial conjecture fails. At this point a question is posed: What is the relationship when P is "between" B and C or "between" A and C?

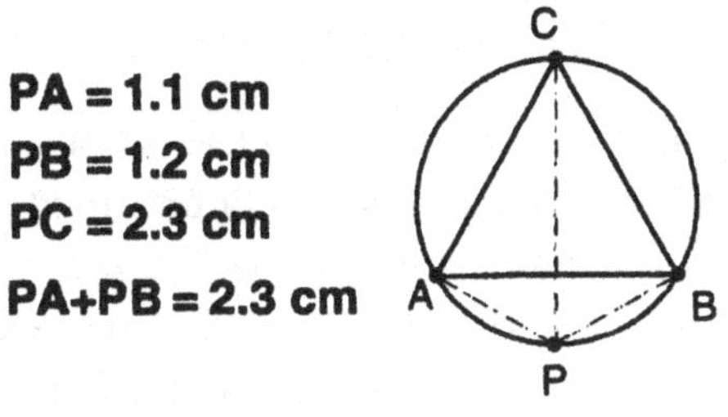

PA = 1.1 cm
PB = 1.2 cm
PC = 2.3 cm
PA+PB = 2.3 cm

Figure 7. PC = PA + PB

PA = 2.2 cm
PB = 0.9 cm
PC = 1.4 cm
PA+PB = 3.1 cm
PB+PC = 2.2 cm

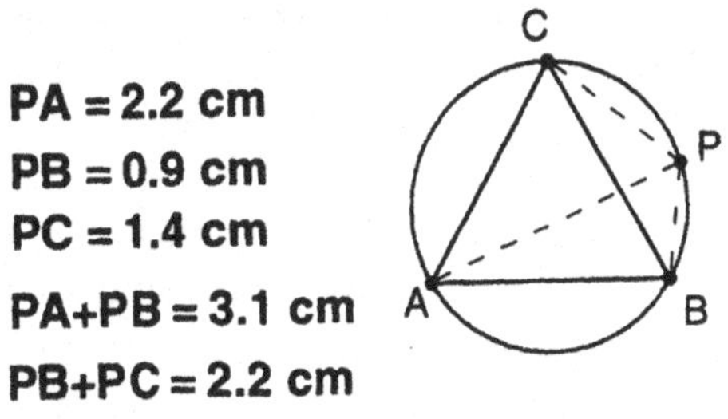

Figure 8. PC ≠ PA + PB but PA = PB + PC

As learners examine Figure 8, they may realize that PA = PB + PC when P is between B and C and when P = B or P = C. Additional dragging supports the conjecture. As learners continue dragging point P, they observe that PB = PA + PC for the case when P is between A and C and when P = A or P = C. To test the conjecture for other equilateral triangles, they drag any flexible point of △ABC (i.e., vertex A or B) and test each conjecture. They observe that the three conjectures seem to hold for other equilateral triangles. From all the observations, learners can state the following generalization for any equilateral triangle:

> Let P be any point on the circumcircle of equilateral △ABC. If P is on the minor arc AB, then PC = PA + PB. If P is on the minor arc BC, then PA = PB + PC. If P is on the minor arc AC, then PB = PA + PC. In other words, the distance from any point on the circumcircle of an equilateral triangle to the remotest vertex is equal to the sum of its distances to the other two vertices.

In this example there are three critical cases to investigate, namely when P is on each of the three minor arcs determined by the vertices of the triangle. The conjecture also holds for the endpoints of the arcs (i.e., the vertices of the triangle), which are extreme cases; a particular conjecture (e.g., PC = PA + PB) holds for these points but does not hold for points beyond the endpoints of the arc.

Viviani's Problem[2]

Our last example in which a lack of systematic dragging affects the conjecture that learners formulate is Viviani's problem:

> Let △ABC be an equilateral triangle. Find all the points for which the sum of their distances to each side of the triangle is as small as possible.

Figure 9 (p. 97) displays the geometric configuration of Viviani's problem. Learners drag point P until they find an unexpected solution: Any point of the interior of an equilateral triangle seems to minimize the sum of its distances from each side of the triangle (Figure 10, p. 98).

---

[2] Vincenzo Vivani (1622-1703) was a Florentine mathematician who studied under Galileo.

Further dragging of point P and any of the flexible vertices of the equilateral triangle seems to support the initial conjecture. A natural question arises, "Are there other points that satisfy the optimal requirement?" As learners drag point P from the interior to the exterior of the triangle, they may realize that any point on the sides of the equilateral triangle also satisfies the minimum criterion (Figure 11, p. 98). To verify that the sum of the distances of any point on the equilateral triangle to the sides of the triangle is minimum, they construct the configuration by constructing the interior of the triangle and then merging the point P to the triangle. Dragging point P along the sides of the triangle confirms the conjecture. Finally, dragging one of the flexible vertices of the equilateral triangle and P verifies that this conjecture seems to be true for other equilateral triangles. In summary, the following conjecture solves Viviani's problem:

> The set of points for which the sum of the distances to the sides of an equilateral triangle is minimal is the triangle and its interior.

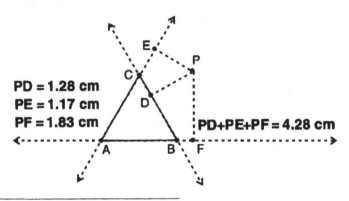

Figure 9. Configuration to investigate Viviani's problem

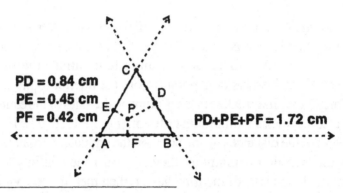

Figure 10. Finding a solution to Viviani's problem

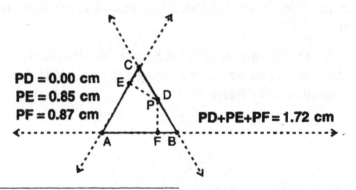

Figure 11. Finding another critical case to solve Viviani's problem

In this example, there are three sets of critical cases to find and test: the exterior of the triangle, the interior of the triangle, and the sides of the triangle. The sides of the triangle are extreme cases because points "beyond" the sides of the triangle fail to satisfy the minimum criterion.

Reflection and Concluding Remarks

We have used these problems in a variety of formats: individual, small group, or whole-class investigations, homework assignments, or as assessment items with a diverse teacher population, including elementary and secondary mathematics teachers at both the preservice and inservice levels. In all of these examples, some learners arrived at incomplete conjectures due, at least in part, to their failure to drag systematically to find and test critical cases. In fact, the authors arrived at an incomplete conjecture when investigating Viviani's problem for the first time. All of us (i.e., teacher educators, teachers, and students) need to learn to perform non-routine investigations within IGEs.

As geometry teacher educators, we constantly look for challenging geometric problems that can be investigated using IGEs to enhance our abilities to formulate mathematical conjectures, an essential aspect of doing mathematics. We consider this to be an important aspect of our professional development. Furthermore, we also need to help teachers develop their ability to make conjectures. We believe that an essential activity of doing mathematics is investigating challenging problems to formulate conjectures during geometry content courses, inservice workshops, and professional development seminars. Such activities will help teachers prepare to provide their own students with opportunities to make geometry conjectures (e.g., NCTM, 2000).

IGEs facilitate the investigation of complex mathematical conjectures. Indeed, it would be difficult to investigate the problems described in this paper without the support of IGEs. However, as discussed earlier, learners need content knowledge about construction, dragging, and critical cases to formulate complete conjectures within IGEs. We have noticed that most learners come to understand the difference between constructing and drawing as well as between dragable and non-dragable points with a few examples. However, learning to drag systematically to find and test all critical cases of a challenging problem before formulating a conjecture is a complex process.

This knowledge about issues related to dragging helps teacher educators and teachers design effective activities to help their students develop their skills investigating conjectures. As in problem solving, there are not step-by-step procedures to guide the formulation of a conjecture. So, we develop geometric tasks in which learners are encouraged to drag one of the flexible points horizontally, vertically, or in circular form until they find a counterexample for the pattern stated in the initial conjecture or until they find additional plausible instances that satisfy the conditions of the given exploration. In most cases, this hint helps them find and test a critical case.

When working with secondary mathematics teachers, the importance of developing students' abilities to formulate geometry conjectures is an essential activity of doing mathematics; teachers discuss and reflect on the power of IGEs to help in investigating or solving challenging problems (NCTM, 2000). In conclusion, IGEs facilitate the investigation of conjectures involving several cases. However, teacher educators, teachers, and students need to learn how to drag systematically to find and test all the critical cases of a conjecture. Otherwise, they may formulate incomplete conjectures. Dragging systematically within interactive geometry environments pays off. It is not a drag!

References

Hölzl, R. (1996). How does "dragging" affect the learning of geometry? *International Journal of Computers for Mathematical Learning, 1,* 169-187.

Hoyles, C., & Noss, R. (1994). Dynamic geometry environments: What is the point? *Mathematics Teacher, 87,* 716-717.

Jackiw, N. (2001). The Geometer's Sketchpad, Ver. 4.0. Berkeley, CA: Key Curriculum Press. Software.

National Council of Teachers of Mathematics. (2000). *Principles and standards for school mathematics.* Reston, VA: Author.

Polya, G. (1954). *Mathematics and plausible reasoning: Induction and analogy in mathematics.* Princeton, NJ: Princeton University Press.

Texas Instruments. (1998). Cabri Geometry II. Dallas, TX: Texas Instruments. Software.

José N. Contreras, Associate Professor of Mathematics at The University of Southern Mississippi, teaches mathematics courses for both elementary and secondary mathematics teachers. He earned a doctorate degree in mathematics education from The Ohio State University in 1997. Prior to coming to the United States, he was a high school teacher in Mexico for seven years. He is interested in integrating problem posing, problem solving, conjecturing, technology, and realistic mathematics education in teaching and teacher education. He has served on the editorial panel of *Teaching Children Mathematics*. [Jose.Contreras@usm.edu]

Armando M. Martínez-Cruz, Professor of Mathematics at California State University, Fullerton, received his PhD in mathematics education in 1993 from The Ohio State University. Since then he has been a faculty member at the National University of Mexico in Mexico City and Northern Arizona University. His professional interests include solving and posing problems, especially within technology environments, and nurturing teachers' dispositions in these mathematical processes. [amartinez-cruz@Exchange.FULLERTON.EDU]

Sharp, J.
AMTE Monograph 1
The Work of Mathematics Teacher Educators
©2004, pp. 103-118

7

Spherical Geometry as a Professional Development Context for K-12 Mathematics Teachers

Janet Sharp
Montana State University

This chapter describes a professional development experience in which teachers explored spherical geometry en route to developing knowledge about the van Hiele learning theory for geometry. The stage theory describes identifiable learner actions representative of various levels. During the session described here, teachers became aware of their own learning and thinking, classifying moments according to the theory levels. A critical piece of the professional development experience was teachers' simultaneous growth in geometry knowledge. Through growth in geometry they became convinced that the learning theory actually corresponded to the process of learning.

We all tend to teach like we were taught (Lortie, 1975; Russell, 1997; Schifter, 1997). But many K-12 mathematics teachers are making efforts to teach differently than they were taught. Mathematics teacher educators, therefore, must share the responsibility to provide opportunities for those teachers to *learn* in a different way. At the same time, just learning new and more content is not enough. Teachers must also analyze learning theories and connect those theories to practice. This paper describes a professional development experience for a group of practicing K-12 teachers in which spherical geometry served as new *content* and as a *context* for demonstrating the van Hiele theory of geometry learning.

Connecting Learning Theory to Teaching Practice

Many practicing teachers perceive a gulf between theory and practice. As teacher educators, we can *and should* build bridges across this gulf by guiding teachers' development of mathematical knowledge while simultaneously augmenting their knowledge about learning theories. Layering mathematical learning between discussions about an associated learning theory certifies to teachers that the learning theory *works* because they recognize it is accurately explaining their own learning. This kind of connected professional development sets the stage for teachers to examine worthwhile mathematical tasks from both teacher and student points of view (National Council of Teachers of Mathematics [NCTM], 1991). A worthwhile task is one that challenges teachers' mathematics knowledge. Using activities in which the experienced teacher immediately knows the result or approach may demonstrate how the learning theory is *supposed* to work. It is not, however, as powerful as affirming that it *does* work, which occurs when teachers comprehend aspects of a learning theory as a result of engagement in a mathematical task. Studying challenging mathematical tasks in this way encourages teachers to revise assumptions about how mathematics is learned.

Connecting Mathematics to Learning Theory

Challenging tasks can be created from spherical geometry because many teachers experience limited collegiate preparation in non-Euclidean geometry. At the same time, all mathematics teachers know some geometry because of the spatial thinking required in all areas (e.g., graphs, diagrams, polygons, and number images). They have some existing geometric knowledge on which to hook new learning. As teachers encounter unfamiliar geometry concepts, they are able to analyze their own learning. Teachers can recognize how the van Hiele theory of geometry knowledge would allow them to identify students' thinking in sequence, to frame questions, and to select tasks that match their learning needs (Crowley, 1987; Fuys, Geddes & Tischler, 1988; van Hiele, 1986). If instruction is at a level different from the level of the students, teachers and students will not understand each other

(Senk, 1989; Teppo, 1991). Moreover, geometry knowledge does not just naturally develop with age. Progressing through the van Hiele levels results only from focused and purposeful instruction (Crowley, 1987).

A Brief Summary of van Hiele's Theory

The van Hiele theory describes five successive levels of geometric thinking. This stage theory describes learners progressing from one level to the next for each geometry idea learned. Levels do not necessarily correspond to chronological ages (Burger & Shaughnessy, 1986; Teppo, 1991). Rather, levels identify how a learner is thinking about a particular geometry idea at a particular time. A learner could even be at different levels for different geometry ideas (Burger & Shaughnessy, 1986; Teppo, 1991). If teachers study the van Hiele theory while constructing (or re-constructing) geometry knowledge, they can appreciate nuances of geometry and geometry learning they may not have known before (Swafford, Jones, & Thornton, 1997). This paper begins with a summary of van Hiele's theory before describing a few tasks from spherical geometry used to develop teachers' knowledge.

Visualization – level 0

At level 0, learners create a mental imprint of the whole shape or concept in question. At this time, they do not consider individual components of the object. They simply check for a match between the physical item or picture and their mental imprint, not really noticing attributes, such as the number of sides of a polygon.

Analysis – level 1

At level 1, learners leave behind holistic visualizations and begin to pick shapes apart. They step close to the object to study components. They might make a list of properties, take several measurements, or identify other empirical features.

Informal Deduction – level 2

At level 2, learners realize some fact as a consequence of something already known about the situation. Here, learners create and use definitions easily. They can follow someone else's basic proofs but do not offer their own proofs of their conclusions.

Formal Deduction – level 3

At level 3, learners make and prove conjectures. They use definitions and other previously proven statements in their proofs. Formal deductions are different from informal deductions in that learners make *and support* their claims.

With so many possible levels among a classroom of students and across topics, high school and kindergarten teachers alike should be familiar with levels 0, 1, 2, *and* 3. It is critical for students to reach level 2 before entering high school geometry (Senk, 1989), which contains numerous level 3 expectations.

Rigor – level 4

At level 4, learners rigorously compare different axiomatic systems. This is the level at which college mathematics majors are encouraged to think about geometry.

Progressing Through a Worthwhile
Spherical Geometry Situation

What follows is a description of a half-day professional development experience designed to incorporate information about the van Hiele theory with geometry learning. Twenty-five mathematics teachers, including four elementary teachers who classified themselves as mathematics specialists, participated in the experience with spherical geometry.

The session wove together van Hiele learning theory with spherical geometry using teachers' discussions about their learning. These discussions served as the substance of the session. Teachers routinely recorded reasons behind their mathematical comments on the board and later categorized them according to van Hiele levels. This process was important because it enabled the teachers to capture comments needed to illuminate and agree on mathematical language and helped the teachers to reveal the appropriate van Hiele theory shortly after each new geometry topic was discussed. Comments from Karen (5th), Gary (8th), Corrine (6th), and Al (10th) seemed to characterize overall goals of the session. Their thoughts and descriptions are included throughout as examples of teachers' approaches to the activities.

Ascertain Existing Knowledge

The session began with teachers drawing sketches of and defining line, segment, ray, circle, polygon, triangle, and parallel and perpendicular lines on the plane. This opening discussion developed a common understanding of the undefined terms, line and point, so teachers could easily use and communicate about these ideas. Once a common understanding was established, the teachers explored consequences of transferring such figures from the familiar plane to the new surface of a sphere. *Would familiar objects, such as a triangle, look the same or have the same properties when drawn and analyzed on a sphere instead of a plane?*

Teachers discussed their images of *point* and reached consensus on *a location*. Some teachers wanted to add *on the plane*; others wanted to add *in space*. Both ideas developed a more general image. For *segment*, teachers settled on "the set of all points on a straight path between two (end)points." Then teachers described *line* as "a straight extension of a segment in both directions." Engaging in a discussion of *straight* versus *level* clarified subsequent discussions of lines on the sphere. The group initiated this discussion by asking, "Is a car traveling uphill from point A to point B going *straight* even though it is not going *level*?" It was important for uphill to be straight, even though uphill is not level. Equipped with agreed-upon definitions and specific mathematical language, teachers were ready to begin explorations on a sphere.

Develop Level 0, Visualizing Shapes on the Sphere

In pairs, teachers worked with plastic playground balls, string, and washable markers. They drew and labeled two points, A and B, on the ball, fairly far apart. Then, they stretched string between the two points and drew along that string to build a segment $\overline{AB}$ based on their earlier description. They extended the segment in both directions to build a line and immediately noticed their extensions met to create a circle. Asking the teachers to describe how this image contradicted their existing visualizations of line or segment again resulted in a record of teachers' reasons on the board. As a group, we separated level 0 phrases (e.g., "That's how a line looks from above" or "That's not how a line looks,") from level 1 phrases (e.g., "Well, it's straight, it doesn't curve, it's a straight path" or "it has two endpoints") and from level 2 phrases (e.g., "But wait, since it's a straight path between A & B, it has to be a segment").

Teachers studied the comments and classified them into van Hiele levels. Naturally, identifying levels was never intended to disparage anyone's level of thinking; rather it served to illustrate how learners *really do* progress serially, regardless of their age, for *any* new topic learned. Thus, everyone begins at level 0, even adults. Recognizing the different kinds of comments helped teachers understand how students may be at different levels for different concepts. That is, students might be thinking at level 0 when thinking about lines, but be thinking at level 1 when thinking about segments. Teachers remembered comments from their classrooms that represented thinking at level 0. Karen recalled a student who studied a plane diagram, and had erroneously assumed one of the angles was 90°, simply because it looked that way. Corrine remembered a student who thought two rectangles were not congruent because one looked larger than the other one. Gary remembered a student struggling with similarity ideas and commenting that two triangles were similar because they looked the same. Al remembered hastily drawing two triangles on the board; his students hesitated to apply the side-angle-side theorem because his drawings did not look congruent.

Develop Level 1, Analyzing Figures on the Sphere

Layering mathematics with learning theory continued. Before developing a deeper awareness of level 1 thinking, teachers returned to mathematics and drew lines, segments, and circles on their spheres (the playground balls). It was important to look at their drawings (level 0) to recognize properties and non-properties (level 1) that could be associated with each idea they generated, with the focus on those properties that highlight thinking at level 1. Karen pointed out that a property of a line segment is that it has two endpoints. The consummate teacher, she put the idea into *teacher* words, "If a bug is sitting on an orange at point A and some bug food is sitting on the orange at point B, the bug would walk straight from one point to the other point." At level 1, learners recognize properties by measuring and comparing parts of figures. She additionally noted, "If it was a smart bug, it would take the shorter of the two paths between the points!" Thus, she noted another spherical property: *Between any two points there are exactly two line segments*. (See Figure 1.)

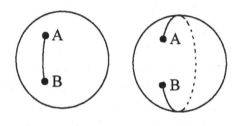

Figure 1. The two segments between A and B

At this point, Karen was on the precipice of a level 2 informal deduction. She started with an observed property of a line segment (it has two end points) and claimed that between any two points there will be two distinct line segments. Learners, like Karen, often move back and forth between levels as they explore and solidify their thinking. Properties *are* a consequence of a figure's characteristics. But, knowing a property as an informal deduction (level 2) is different from identifying the property (level 1) through measurement or observation. As Karen thought further about the existence of two segments for any pair of points, she drew several more pairs of points and associated segments. Thus, empirical evidence served as her reasoning, making her reasoning at level 1 (analysis). Had Karen reasoned, "Since any two distinct points are contained in exactly one (unique) line, those points separate it into exactly two segments," her claim would have been at level 2 (informal deduction). *(I feel compelled to reiterate. Sharing teachers' insights is not intended to be disrespectful, but to illuminate theory and to admire teachers, like Karen, who make the journey.)*

Quick Trip to Level 2: Making Definitions for Lines and Segments

As anticipated, teachers wanted to know more about deductive reasoning. Some circles drawn earlier encompassed the girth of the sphere; others were small islands. As a group, we sorted the balls into these two classifications to highlight the specialness of circles traveling the girth and developed a shared definition. *Girth circles are the collection of points that fall along the longest straight path around the ball*; these circles are known as great circles. Teachers quickly noted that girth circles were exactly like lines. It was vital to merge these two ideas and redefine line as a great circle. Teachers compared level 2 reasons (a definition) to level 1 reasons (list of properties) that lines were great circles. Developing a shared definition of segment fell into place: *A continuous portion of a great circle, bounded by two endpoints*. Corrine pondered her great circle, noting a property. "It cut my sphere in half." Her comment allowed us a smooth transition back to level 1 discussion.

Examples of Moving Between Levels 1 and 2

Once the teachers satisfactorily listed properties and definitions for line, segment, and circle, it was time to visit (1) triangles, (2) perpendicular lines, and (3) parallel lines.

Triangles. Teachers drew triangles on their spheres, with prompts to make large triangles. Small drawings more closely approximate plane drawings, which fail to stress spherical properties. Teachers discussed whether or not the shapes *looked like* 3-sided polygons and commented about the importance of level 0 reasoning to make initial statements. As teachers worked, it proved valuable to again separate *properties* (e.g., segments, closed, simple) from *visualizations* so teachers could compare their responses.

Al's unusual triangle, shown in Figure 2, initiated discussion of *definition*. Several teachers deduced, "If the definition of triangle is a *3-sided polygon*, then whether or not the figure is a polygon partly determines whether or not the figure is a triangle." Their definition of polygon was, "closed, simple figure composed only of segments." This led them to recall their definition of segment as a continuous portion of a great circle bounded by two endpoints. The original definition did not include anything about the *shortest* distance. Al's figure was indeed composed of three segments, and hence, was a triangle. Even though the segment traversing the back of the sphere clearly did not travel the shortest path and severely violated their commonly-held visualizations (level 0) of triangles, the teachers eventually abandoned their visualizations and stood by their original definition (level 2) of a triangle. This process demonstrated an example of movement to level 2. The teachers experienced how deductions could be culled from properties.

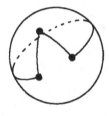

Figure 2. Al's unusual triangle

Al's unusual triangle prompted Gary to make the following deduction: "Since any two [distinct] points can create two different segments, then any three [distinct] points would not correspond to exactly one triangle." Gary openly struggled with his knowledge of triangle. On the plane a collection of three distinct points created exactly one unique triangle. But, on the sphere, this would not be the case. He drew some examples to support his claim. See Figure 3.

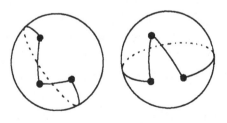

Figure 3. Two of Gary's triangles

Even though Gary had just learned to ignore the concept of shortest distance for his segments, general ideas about measurement should not be ignored. Measurement provides important evidence to level 1 thinkers. Learners see a trend in a collection of measurements as they examine empirical evidence. In much the same way, learners can recognize other trends in a collection of examples, which still represents level 1 reasoning. Corrine distinguished these ideas, "So, if Gary had drawn a bunch of triangles [using the same three points] to decide more than one triangle existed, that would be level 1. But, since he said it was because of his definition of segment, it was level 2 thinking."

By now, many teachers did not trust their intuition and wondered whether or not length or degrees could be measured. A string and ruler to measure length and a flexible, plastic protractor to measure degrees were good tools for exploration. Only one reminder was needed. Length measurement must travel along the surface of the sphere and not tunnel through the ground. For example, what path would an airplane take to fly from Dallas to Moscow? The flight path would not tunnel through the earth, but would follow the curvature of the Earth. One teacher remembered a recent trip when a flight attendant called his flight a "segment."

Teachers were relieved to find little difference between measuring an angle on the sphere and the more familiar process of measuring planar angles. They set the protractor into its proper position (zero at the angle's vertex) and curved the plastic along one leg of the angle indicating the initial side. Then, they estimated the degree of openness from the other leg of the angle. The concept was clear from this rudimentary procedure. By analyzing empirical angle measurements (level 1), teachers discovered the sum of the angles in a triangle is always greater than 180°. Al deduced (level 2) that the angle sum of a quadrilateral would always be greater than 360°: "Any quadrilateral can be cut into two triangles and each of those triangles will have an angle sum greater than 180°. More-than-180 plus more-than-180 is more-than-360."

Perpendicular Lines. Perpendicular lines can be built from the creation of a 90° angle. After asking teachers to visualize a sketch of perpendicular lines, they returned to the sphere. Corrine pointed out, "The intersection of any longitude and the equator forms a 90-degree angle. Hey, so does any longitude and any latitude!" After a moment's hesitation, she quickly corrected herself, "Oh wait! Those other latitudes [besides the equator] aren't lines, so that's not right!" Corrine's insight demonstrates nice informal deductive thinking.

Teachers used protractors to draw perpendicular lines on their spheres. Enthusiasm soared over realization that perpendicular lines intersect twice, once in front and once in back, forming 8 right angles. Teachers grew increasingly interested in comparing figures on the sphere to figures on the plane.

Resume Level 2, Drawing Informal Deductions

Parallel Lines. The next task was to draw parallel lines. Keeping with the approach of comparing figures on the plane to corresponding figures on the sphere, the group modified the most common definition of parallel lines from "Two lines, on the same plane, which do not intersect" to "on the same sphere." This definition clearly states three premises for figures to be *parallel lines*: The figures have to be "lines;" they have to be "on the same sphere;" and they must be "non-intersecting."

The request to draw parallel lines resulted in an intense debate, allowing further accentuation of the difference between an informal deduction (level 2) and a property (level 1). Nearly all teachers initially tried to build parallel lines by drawing two equidistant circles. Their drawings appeared to satisfy their visual images (level 0) of parallel lines because these shapes were equidistant and did not intersect. But, as Gary later deduced, "My drawing is like two latitudes and latitudes are not lines. We cannot call two equidistant circles parallel lines because [at least] one of them is not a line in the first place!" Thus, the examples in Figure 4 (p. 114) were not solutions. Continuing to draw two non-intersecting lines led to a problem. No matter how hard they tried, teachers could not meet all three requirements of their definition. At least one premise was always violated. But this was a good dilemma, for it was here that teachers solidified their notion of line. *All lines would intersect and intersect twice.*

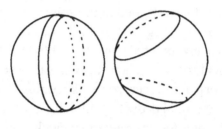

Figure 4. Initial attempts at parallel lines

Struggling with the task to draw lines that never intersected seemed ultimately to authenticate teachers' understanding of the van Hiele theory. The eventual deduction that parallel lines do not exist materialized when teachers discarded all drawings of two non-great circles. Their circles did not intersect, which *was* one of the premises. But, their circles weren't *lines*. The ensuing discussion, with three premises to explore, promoted level 2 thinking and recognition of colleagues' informal deductive thinking. At the end, Karen said, "So equidistance is a *property* of parallel lines on the plane because of our definition. I can draw two shapes equidistant apart on my sphere, but I should not *deduce* the shapes are lines." Al added, "Since all lines are great circles cutting the sphere in half, all lines will intersect." So, what had started as a level 1 reason (trying to draw several lines to validate thinking) in fact ended with a level 2 deduction. Teachers' knowledge had grown and they knew it. They were eager to deal with more premises and explore consequences of those premises on the sphere. Some teachers, whose curiosity had been piqued, developed their own conjectures for the sphere. (See the Appendix.)

Summary

After this professional development experience, teachers could articulate informal deductions about spherical geometry ideas. They accomplished this by comparing plane figures to related spherical figures. Most notably, they related personal classroom scenarios where van Hiele ideas had appeared. Fifteen of the 25 math teachers have incorporated a spherical geometry unit into their teaching. Al said, "I now have a reason to teach, not skip, the section on spherical geometry, it helps them understand the process of deduction." Each spring Corrine teaches some basic ideas from spherical geometry to her sixth-graders and shared her experiences in an article (Sharp & Heimer, 2002.) Gary lamented that he did not learn about van Hiele in his methods class, but said, "I consider myself a life long learner. At least I've learned about it now. I think I'm a better teacher because of it." He recognized the importance of how a shape *looks* to students at first. Karen uses spherical geometry for enrichment with fifth-graders, being careful to focus on developing informal deductions. Like Corrine, Karen published a paper (Sharp & Hoiberg, 2001). Karen also gave a NCTM presentation (Hoiberg, 2001), which was highlighted in the March 2001 NCTM newsletter announcing the conference. When excellent teachers share their efforts to implement theory, they bring credibility that other teachers appreciate.

As a general guideline, professional development must resonate with the practical knowledge of classroom teachers. At the opening of this session, teachers voiced skepticism at studying theory, presumably because they had not seen connections to their classrooms in the past. But disbelief soon disappeared. The van Hiele theory came to life when teachers analyzed their own learning and recounted classroom stories that were consistent with it. Nuances of the theory are known intimately to Karen, Gary, Al, and Corrine, arguably because they learned new mathematics content at the same time and they could recall relevant comments from their own students. Valuing classroom knowledge on an equal footing with theory is ultimately the bridge. For these teachers, a bridge between theory and practice has begun.

References

Burger, W. F., & Shaughnessy, J. M. (1986). Characterizing the van Hiele levels of development in geometry. *Journal for Research in Mathematics Education, 17*(1), 31-48.

Crowley, M. L. (1987). The van Hiele model of the development of geometric thought. In M. M. Lindquist (Ed.), *Learning and teaching geometry K-12, 1987 yearbook of the National Council of Teachers of Mathematics* (pp. 1-16). Reston VA: National Council of Teachers of Mathematics.

Fuys, D., Geddes, D., & Tischler, R. (1988). The van Hiele model of thinking in geometry among adolescents. *Journal for Research in Mathematics Education Monograph #3*. Reston, VA: National Council of Teachers of Mathematics.

Hoiberg, K. (2001, April). *van Hiele and 5th graders – You bet!* Paper presented at the annual National Council of Teachers of Mathematics meeting, Orlando, FL.

Lortie, D. C. (1975). *Schoolteacher*. University of Chicago Press: Chicago.

National Council of Teachers of Mathematics. (1991). *Professional standards for teaching mathematics*. Reston, VA: Author.

Russell, T. (1997). Teaching teachers: How I teach IS the message. In J. Loughran and T. Russell (Eds.), *Teaching about teaching: Purpose, passion and pedagogy in teacher education* (pp. 32-47). London: Falmer Press.

Schifter, D. (1997). *Learning mathematics for teaching: Lessons in/from the domain of fractions*. Newton MA: Education Development Center, Inc. (ERIC Document Reproduction Service No. ED 412 122)

Senk, S. (1989). van Hiele levels and achievement in writing geometry proofs. *Journal for Research in Mathematics Education, 20*, 309-321.

Sharp, J., & Heimer, C. (2002). What happens to geometry on the sphere? *Mathematics Teaching in the Middle School, 8*(4), 182-188.

Sharp, J., & Hoiberg, K. (2001). And then there was Luke: The geometric thinking of a young mathematician. *Teaching Children Mathematics, 7*(7), 432-439.

Swafford, J. O., Jones, G. A., & Thornton, C. A. (1997). Increased knowledge in geometry and instructional practice. *Journal for Research in Mathematics Education, 28*(4), 467-483.

Teppo, A. (1991). van Hiele levels of geometric thought revisited. *Mathematics Teacher, 84*(3), 210-221.

van Hiele, P. (1986). *Structure and insight: A theory of mathematics education.* Orlando, Florida: Academic Press.

Janet Sharp, Associate Professor of Mathematics Education at Montana State University in Bozeman, teaches undergraduate and graduate courses in mathematics education. Her interests include supporting future and practicing teachers' uses of appropriate pedagogy to build and analyze geometry and rational number knowledge in K-12 learners. She is also interested in grounding this teaching and learning in culturally sensitive contexts. [sharp@math.montana.edu]

Appendix
Teachers' conjectures for figures *on a sphere*

1. Every circle has *(two)* unique centers.
2. Every non-great circle on the sphere has *(2 distinct)* radius measures.
3. All radii of a great circle are *(congruent)*.
4. Since any two lines intersect, *(a 2-sided polygon, called a lune, exists)*.
5. Since parallel lines do not exist on the sphere, *(parallelograms)* cannot exist either.
6. The angle sum of a pentagon would be *(greater than 540)*.
7. Any two unique lines subdivide a sphere into *(four)* regions.
8. In a quadrilateral with four equal angles, the diagonals are *(congruent)*.
9. The ratio of circumference ÷ diameter is *(less than π (3.14159...))* for all circles.
10. The *(circumferences)* of all great circles on the same sphere are of equal length.
11. A segment joining two points of a non-great circle will *(not fall)* on that non-great circle.
12. Non-congruent similar triangles *(do not exist)* on the sphere.

Van Zoest, L.R.
AMTE Monograph 1
The Work of Mathematics Teacher Educators
©2004, pp. 119-134

8

Preparing for the Future:
An Early Field Experience that Focuses
on Students' Thinking[1]

Laura R. Van Zoest
Western Michigan University

Preservice teachers' first experiences in their mathematics teacher education programs orient them on the teacher professional learning (TPL) continuum. Rethinking the role of the field experience at the beginning of the TPL continuum led to the development of the Explore-Plan-Teach-Reflect-Teach-Reflect (E-P-TR-TR) sequence, an early field experience that shifts undergraduates' involvement in classrooms from observing teachers and their teaching to engaging with students and their mathematical thinking and understanding. This paper describes one such E-P-TR-TR sequence, shares some of the undergraduates' learning during the process, and provides suggestions for others interested in implementing such a field experience.

As preservice teacher educators, we orient our undergraduates as they embark on the teacher professional learning (TPL) continuum that extends through their entire career as a teacher. Traditionally, undergraduates' first experiences in schools, beyond being students, are to observe classroom teaching. The pitfalls of this approach are well documented (Evans, 1986; Feiman-Nemser & Buchmann, 1985; Goodman, 1985; McDiarmid, 1990; Zeichner, 1985), and the result can be a negation of other aspects of a teacher education program. For example, undergraduates who observe classroom teaching based on a transmittal model of knowledge acquisition may not be able to

---

[1] Throughout, "student" is used when referring to K-12 students and "undergraduate" for university teacher-preparation students.

envision a classroom based on ideas put forth in the National Council of Teachers of Mathematics (NCTM) *Standards* (1989, 1991, 1995, 2000). Even when undergraduates are able to observe classrooms reflective of the *Standards*, it is not clear that they have the tools needed to identify critical aspects of the classroom environment and the teacher's orchestration of those aspects. Or, even if they can identify what the classroom teacher did, they may not be able to translate those actions into their own classroom practice. As a result, we, as teacher educators, must think harder about what would provide an effective beginning to the TPL continuum.

As my departmental colleagues and I engaged in conversations about our teacher education programs, we identified mathematical reasoning as a central focus for achieving our goal of preparing teachers to learn mathematics with understanding, to reflect on what it means to understand mathematics, and to teach for mathematical understanding. As a result of these conversations, we have increased the attention we pay to our undergraduates' thinking about important mathematical ideas and the way in which students come to understand them. Along with changing the way in which we organize and orchestrate our mathematics teacher education courses, this focus has caused us to rethink the role of the field experiences at the beginning of the TPL continuum.

Specifically, we have developed an early field experience that shifts our undergraduates' involvement in classrooms from observing what the teacher is doing to engaging with the students and their mathematical thinking. Our relationship with local teachers allows us to engage our undergraduates in two explore-plan-teach-reflect-teach-reflect (E-P-TR-TR) sequences each semester. In the following, I describe one such E-P-TR-TR sequence, share some of our undergraduates' learning during the process, and provide suggestions for others interested in implementing such a field experience.

Background

The sequence focused on here took place in the first of three mathematics teacher education courses taken by secondary mathematics teacher education majors in our program. The focus of the three-credit course is teaching mathematics in the middle school. The course is designed to reflect the instructional approaches advocated in the *Standards* and makes use of tools such as reform curricular materials, written and video case studies, and sets of student work. One of the goals of the course is to problematize familiar mathematical concepts and procedures so that the undergraduates begin to reconsider some of the mathematics that they thought they knew well.

By carefully choosing problems that can be approached from multiple angles, the instructor can facilitate a discussion and provide undergraduates an opportunity to deepen their mathematical understanding. By orchestrating class discussions that focus on the mathematical reasoning involved in solving these problems, the instructor enables undergraduates to participate in, observe the facilitation of, and discuss their experience with the kind of classroom discussions recommended by the *Standards*. Additional opportunities for learning are created by selecting problems from reform curricula materials; by sharing the origin of the problem after the undergraduates have experienced it themselves, these prospective teachers see for themselves how rich problems can be embedded in a coherent, well-thought-out curriculum. As the undergraduates plan for and teach to small groups of students the same problem they have investigated themselves, they experience a valuable next step in the process by reflecting on how students interact with such problems.

A final written report on the field experience component of the course is worth 10% of the course grade. Other components of the grade include a mathematics autobiography (5%), class preparation and participation (25%), a take-home midterm that focuses on interpreting students' work and connecting it to expectations in the state curriculum framework (25%), and a final exam that includes an analysis of a video clip of classroom teaching (35%).

For the specific E-P-TR-TR sequence described here, the mathematics topic was *median* as a measure of central tendency. This topic was chosen after discussions with the district mathematics coordinator and the teachers involved because it (a) was an area in which students across the district had done poorly on the state assessment, (b) was sequenced within the pacing chart for the days we would be in the 7th grade classrooms, (c) could be adequately addressed in one 45-minute period, and (d) was a topic that the undergraduates thought was simple and straightforward, providing an opportunity for them to reconsider a topic they took for granted. The ideas for the explore and plan phases of the sequence were taken from Investigation 1 in the Connected Mathematics Project book, *Data About Us* (Lappan, Fey, Fitzgerald, Friel, & Phillips, 1996).

There were 18 undergraduates in the course. The schools in which the field experiences took place were part of a medium-size urban district with three middle schools (Grades 7 and 8). During the course of the semester, each undergraduate taught in at least one classroom in each of the three schools.

The Explore-Plan-Teach-Reflect-Teach-Reflect Model

During the *explore phase* of the model, undergraduates engage with the mathematics as learners. They are explicitly instructed not to think of themselves as teachers and not to try to think like middle-school students. This can be difficult, as undergraduates taking a mathematics teacher education course often do not expect to be learning mathematics themselves. The challenge to the instructor is twofold: first, to frame seemingly simple tasks in such a way that they can be analyzed from multiple perspectives; second, to convince the undergraduates that having a deep understanding of a topic like median is as much a part of mathematical learning as finding the median of a list of numbers.

We meet this challenge by maintaining our focus on mathematical reasoning, that is, by continually pushing the undergraduates to justify their solutions or responses — first individually, then in small groups, and finally as a whole class — until they achieve a deep understanding of the topic under discussion. The stage for this approach is set on the first day of class and is reinforced throughout the course. Thus, by the time we begin the field experience about halfway through the semester, the undergraduates understand the expectation to explain the "whys" as well as the "hows" of their solutions.

The *plan phase* can begin with the topic and basic activities already chosen or with the undergraduates determining the activities that should be used. In either case, the bulk of the planning time is spent on discussions about engaging the middle-grades students with the material, questions to ask to help them develop a better understanding of the mathematics, and ways to build on students' responses. Building on students' responses seems to be the most difficult aspect of the assignment for undergraduates with little or no classroom experience, yet it is an essential piece of their development. To facilitate this thinking, the undergraduates brainstorm possible student responses and plan how they will use those responses to achieve the goals for the lesson. (See Stigler, Fernandez, & Yoshida (1996) for an example of how this can be incorporated into a lesson plan.) Prior to the first teaching session, the undergraduates have difficulty anticipating possible student responses and rely on their own experience with the problems. After they have interacted with middle-grades students, their understanding of how students make sense of the problem increases and they are able to anticipate a wider range of responses. For example, during the planning phase undergraduates had difficulty thinking beyond their own correct approach to the problem. However, during the first reflect phase, they talked with authority about how students might think that the size of an added value would affect the median rather than its position relative to the current median. They then strategized teaching actions to address this misconception in a manner that supported students' understanding.

The *teach phase* involves the undergraduates participating in three 45-minute middle-school class sessions in one school. Each group of four students in a classroom is assigned an undergraduate teacher and an undergraduate documenter. The role of the undergraduate teacher is to implement the planned lesson. The role of the documenter is to observe the teaching and document what happens. The classroom teacher is encouraged to observe the different groups as they work and to share his or her observations with the undergraduates as well. This sharing has occurred through individual conversations, written observations given to the university instructor, and, occasionally, through attendance at the reflecting (debriefing) sessions. Individual schedules varied, but each undergraduate spent at least one period teaching and at least one period documenting for each teaching phase.

As a class, we experimented with a variety of documenting forms, ranging from checklists to scripting. The form that seemed to give the most useful information is a simple 3×2 grid with "Giving the right amount of information," "Pushing students to think harder," and "Checking students' understanding" down the side and "Successes" and "Missed opportunities" across the top. Documenters are expected to "Provide specific examples for each box in the grid." Figure 1 shows a fairly typical example of a completed form. At the end of the lesson, the documenters give their documentation sheets to the undergraduate teacher they observed so that he or she can use that information to inform his or her written reflections.

The *reflect phase* includes both a written reflection and a whole-class reflecting session. In preparation for the reflecting (debriefing) session, the undergraduates are required to complete a field experience reflection. This reflection serves several purposes: (a) to highlight the importance of teachers spending time reflecting on their students' understanding, (b) to lay the groundwork for rich discussions, and (c) to provide the classroom teachers with feedback on their students. The prompt for the reflection is given in Figure 2 (p. 126). By having the undergraduates upload their reflections to the course online learning environment, the professor is able to preview their thoughts and adapt the plans for the reflecting session accordingly.

The reflections tend to range from two to five pages in length and vary in quality and level of detail. It seems, though, that the undergraduates take seriously their charge to pay close attention to what happened in the teaching episode and to make sense of it. The following is a brief excerpt from a fairly typical reflection describing a segment of the teaching episode:

	Successes	Missed opportunities
Giving the right amount of information	• After the students identified the middle, you told them it was called the median. • After students demonstrated understanding of median, you asked what effect adding names of different lengths would have on it.	• A girl answered a question incorrectly. You said "no" instead of asking her to explain her thinking.
Pushing students to think harder	• What do you see on the information about Mr. Who's class? Response "there is a line plot." Explain the line plot—good responses. • Can you remove names other than the outside? • Adding names to increase median resulted in a lot of thinking.	• Tony mentioned median from some class, could ask him more about it. • Give more time for students to think. • Ask if adding really long names changed the median more than if the name was just longer than the median.
Checking students' understanding	• What do you mean by that? — getting them to explain • How did you get that? • Asked if names of different lengths could get same results. • What would happen to the median if we added a name with 1018 letters? The students understood that it would not change the median more than one with 16 letters.	• Don't accept "I don't know." Push them to think about what they do know.

Figure 1. Completed documenting grid

Field Experience Reflection

After the teaching session* you should write a reflection in preparation for the reflecting part of each Teach-Reflect cycle. These are the things that should be included in your analysis:

- What the students understood (cite evidence)
- What struggles the students had (cite evidence)
- The approach(es) that the students took to solve the problem
- Actions you took to help the students' understanding
- The students' responses to your teaching actions
- Other observations that you made about the process

Identify the teacher of the class in which you worked and the first names of the students. Submit the written part of your analysis electronically by midnight prior to the debriefing session. Bring a hard copy of your analysis and all supporting documents (student work, etc.) to the debriefing meeting in a form that can be given to the classroom teacher.

*A teaching session is defined as the time you spend with one group from one class. Some of you will participate in multiple teaching sessions during a given teaching cycle. You are encouraged to reflect on each session, but are only required to reflect on one. One of the requirements of your final field experience report will be to reflect on what you learned across all your teaching sessions; thus, it will be useful to keep notes about teaching sessions for which you don't write formal reflections.

Figure 2. Field experience reflection directions

The difficulties began once I started asking them to remove or add two names [to the given list of names ordered by length] to change the median. Each student tried several times to change the median by removing names, both making it bigger and smaller, but could not find a way to do it. They kept taking names away from the top half and the bottom half, and the median continued to stay the same. Then [student name] and [student name] decided to take two names away from only one half of the data, and realized that you could change the median by doing this. Before I was done asking them why they had done that, the girls realized the same thing. I proceeded by asking them to change the median in various ways, and they continued to do well by adding and subtracting names to do this.

In a later excerpt, this same undergraduate analyzed his actions:

To help the students' understanding, I did a few things. I asked them to explain everything they said to me, no matter whether it was a "yes" or "no" response. In addition, I asked them if names of different lengths could affect the data set and its statistical properties, and if so, how it affected them. By doing each of these, I made the students express their thoughts and show me that they knew the concepts that we had been discussing and exploring.

The reflecting session focuses on making sense of the undergraduates' experiences and what these experiences help them understand about teaching and learning in general. In particular, we discuss what they have learned about students' understanding and the connections between what they did as teachers and the students' thinking.

What has been particularly effective is for the undergraduates to be able to repeat the teach-reflect phases of the sequence. Thus, in the instance reported here, the reflect phase was followed by another teach phase with the same lesson in a different school. The undergraduates adjust the lesson plan based on information from the first attempt and have an opportunity to implement some of the knowledge they gained from the first teach-reflect phase. The second teaching is followed by another written reflection and a reflecting session that looks at new learning and compares the two experiences.

We attempt two full E-P-TR-TR sequences each semester. At the end of the two sequences, the undergraduates complete a field experience report. This report utilizes the earlier field experience reflections, but serves a different purpose. Instead of focusing primarily on the students and their learning, the report focuses on the undergraduates and their learning. Undergraduates are first asked to read O'Reilly's (1996) chapter, "Understanding Teaching/Teaching for Understanding," to set the stage for the type of self-analytical thinking that we expect. This reading also reinforces the message that we are all learning about the process of teaching for understanding and should be able to find both things on which we are making good progress and things that need more effort. The undergraduates are then asked to think about their development as a teacher and to write about what they learned about themselves as a result of the field experience. Within their report, they are expected to:

- Demonstrate a high level of self-awareness and insight into students' understanding;
- Provide detailed support for the analysis, including classroom examples that illustrate claims; and
- Make connections to readings from class, including citations in the form (Author, Year) with reference list.

Observation notes that the professor makes about each undergraduate from the teaching sessions, as well as written documentation from the classroom teachers and the documenters, are kept in mind during the evaluation of the reports. Although it is not possible to have complete sets of data on any individual undergraduate, there is enough information to provide a sense of the relationship between the undergraduates' self-perceptions and those of observers.

To provide more concrete support for the evaluations and to increase the depth of the undergraduates' analyses, we are currently piloting the process of having the undergraduates audiotape their teaching episodes, listen to the tapes, and include pieces of dialogue in their written reflections. Although this pilot group has not yet submitted their field experience reports, the discussions they have been having as a result of listening to their tapes suggest taping the teaching episodes may be an important addition to the experience.

Undergraduates' Learning

The goal of the course in which the E-P-TR-TR sequence was embedded is to prepare undergraduates to become teachers able to teach in a manner reflective of the *Standards*. To us, this means a focus on students' thinking and understanding, and on teaching strategies that support this focus. Most of our undergraduates are convinced that these are laudable goals, but few have experienced this kind of teaching prior to this course. Thus, implementing the ideas requires intentional self-monitoring and control against intuitive reactions based on years of learning mathematics more procedurally. As one undergraduate stated in her field experience report:

> It is difficult to know how to proceed when the answers given are correct. My first instinct is to reaffirm the student and move on. However, this places myself as the mathematical authority in the classroom. For instance, when we were writing down definitions for mode and median, the students looked to me for confirmation that the definition was correct. Instead, I should have directed the confirmation back to the data we had just worked with to see if the definition fit with what we had observed. Referring to things besides myself allows the students to evaluate whether or not an answer is correct instead of me telling them.

Almost all of the undergraduates identified instances such as this where their natural tendency was to use an instructional approach that they recognized as not effective.

When the undergraduates were able to implement ideas that they had learned in class, they often were surprised at the outcome: "When I asked [student name] to explain how he took the median out of the class [sic], the students listened to him, then tried it themselves, and I thought, wow, this is actually working." The undergraduates often felt tension between what they thought they were supposed to be doing based on our planning discussions and their conflicting beliefs. One undergraduate alluded to the tension she experienced in the following excerpt from her field experience report:

> I asked if the placement of the numbers or the value of the numbers was more important, and [student name] replied that it was definitely the value of the numbers. So I proceeded to ask him to explain why that was so. He kept telling me that the huge numbers will make the median very large. I didn't want to say no, but that was what I felt like saying. I ended up asking him to add a huge number into the number line [a written out ordered list of the length of the names] and then tell me where it took the median. With the help of the manipulative he was able to change his own answer before I had to say "no," but it is hard to judge if it is more helpful or harmful to let them figure out why on their own. This is where for a person who was not taught in a constructivist classroom, it is hard to teach in a constructivist way. I learned that this is a necessary process, however. If the students come up with it on their own, I can say, "What did you find out?" instead of "What did you memorize about this situation?"

Although it is clear from the undergraduates' writings that they are struggling with some of the ideas presented in the class and with how they will implement them in their future teaching, the E-P-TR-TR sequences do seem to be helping us reach our goals for the course. This is particularly true in comparison to semesters in which the field experience component of the course was a set number of hours spent observing in middle-school classrooms. Those situations often worked at cross-purposes with our goals as the undergraduates typically saw a disconnect between class discussions and what they observed in the schools. The E-P-TR-TR sequences make real the course ideas, and, at least for most undergraduates, demonstrate that the ideas have value in real classrooms.

Suggestions for Implementation

The course discussed here was modified to permit the embedding of a field experience. The three-credit course met for 110 minutes twice a week for 15 weeks instead of the expected 75 minutes. This schedule allowed us to maintain the in-class instruction time even though the undergraduates were out of the university classroom and in the middle-school classroom for four class periods corresponding to the teach phases of two complete E-P-TR-TR sequences. We spent approximately one 110-minute class period for each phase of the E-P-TR-TR sequence. Although this may seem like a lot of time for each phase, the phases serve as launching points to revisit many of the ideas introduced earlier in the course and to foreshadow others that will be covered in more depth later. The explore phase, in particular, has been an important investment of time. Through this phase, undergraduates are stretched in their mathematical thinking, but are able to discuss what instructors must do to lay the groundwork for the kind of learning that they experienced. Classroom norms, questioning strategies, and the ways in which teachers build on students' thinking can all be discussed in the context of a shared experience. Without such experiences and explicit discussion of them, it does not seem likely that undergraduates will be well prepared to implement the expectations of the *Standards*.

The level of intensity — teaching, documenting, the university professor's presence in the schools moving between the classrooms in which the undergraduates were teaching — seemed to contribute to the depth of the discussions in the reflecting sessions. Although it is difficult to identify the impact of any particular component in isolation, it does seem to be critical that the university professor is present in the schools. Focusing the field experience in a few middle-school classrooms and during a compressed period of time made it possible for the professor to be with the undergraduates during their participation in the schools. As a result, the professor could provide specific examples of situations that occurred across teaching groups and could ask more informed questions about the undergraduates' experiences.[2]

A key aspect of the E-P-TR-TR sequence that distinguishes it from other field experiences is the communal nature of the teaching. In any one teach phase, all of the undergraduates of the class are in the same school, and they are divided among no more than three classrooms. This shared experience contributes to the concreteness of the reflecting discussions and allows the group to get beyond using the same language to represent very different realities. The role of documenter strengthens the reflections because the documenter and the undergraduate teacher provide two views of a teaching episode, facilitating both "reflection-in-action" and "reflection-on-action" perspectives (Schön, 1983).

Conclusion

My colleagues and I have found the E-P-TR-TR sequence to be an effective instructional approach. Incorporating the sequence into a mathematics teacher education course has allowed us to capitalize on the benefits of field experiences while avoiding many of the pitfalls (Feiman-Nemser & Buchmann, 1985). In particular, the way in which we have orchestrated the sequence shifts our undergraduates' focus to engaging with students and their thinking and understanding, rather than observing teachers and their teaching. We believe that this is an important first step in preparing teachers who are able to achieve the vision and promise of the *Standards* in their classrooms and an effective beginning to the TPL continuum.

---

[2] We used three teachers' classrooms for each teaching session and spent a total of four mornings in middle-school classes during the semester.

References

Evans, H. L. (1986). How do early field experiences influence the student teacher? *Journal of Education for Teaching, 12*(1), 35-46.

Feiman-Nemser, S., & Buchmann, M. (1985). Pitfalls of experience in teacher preparation. *Teachers College Record, 87*(1), 53-65.

Goodman, J. (1985). What students learn from early field experiences: A case study and critical analysis. *Journal of Teacher Education, 36*(6), 42-48.

Lappan, G., Fey, J. T., Fitzgerald, W. M., Friel, S. N., & Phillips, E. D. (1996). *Data about us*. Palo Alto, CA: Dale Seymour.

McDiarmid, G. W. (1990). Challenging prospective teachers' beliefs during early field experience: A quixotic undertaking? *Journal of Teacher Education, 41*, 12-20.

National Council of Teachers of Mathematics. (1989). *Curriculum and evaluation standards for school mathematics*. Reston, VA: Author.

National Council of Teachers of Mathematics. (1991). *Professional standards for teaching mathematics*. Reston, VA: Author.

National Council of Teachers of Mathematics. (1995). *Assessment standards for school mathematics*. Reston, VA: Author.

National Council of Teachers of Mathematics. (2000). *Principles and standards for school mathematics*. Reston, VA: Author.

O'Reilly, A. M. (1996). Understanding teaching/teaching for understanding. In D. Schifter (Ed.), *What's happening in math class? Reconstructing professional identities* (Vol. 2, pp. 65-74). New York: Teachers College Press.

Schön, D. A. (1983). *The reflective practitioner: How professionals think in action*. New York: Basic Books.

Stigler, J. W., Fernandez, C., & Yoshida, M. (1996). Traditions of school mathematics in Japanese and American elementary classrooms. In L. P. Steffe, P. Nesher, P. Cobb, G. A. Goldin, & B. Greer (Eds.), *Theories of mathematical learning* (pp. 149-175). New Jersey: Lawrence Erlbaum.

Zeichner, K. M. (1985). The ecology of field experience: Toward an understanding of the role of field experiences in teacher development. *Journal of Research and Development in Education, 18*, 44-52.

Laura R. Van Zoest, Associate Professor of Mathematics Education at Western Michigan University, has consistently focused on improving mathematics teaching for secondary school students. During the past ten years, she has worked with inservice middle- and high-school mathematics teachers through several National Science Foundation-funded projects and, whenever possible, has engaged preservice teachers with inservice teachers in project activities. Her research questions revolve around how one becomes an effective teacher of mathematics and the role of learning communities in that process. Recently, she has returned to her original interest in early field experiences and the way in which they set the stage for future learning on the Teacher Professional Learning continuum. [laura.vanzoest@wmich.edu]

D'Ambrosio, B. S.
AMTE Monograph 1
The Work of Mathematics Teacher Educators
©2004, pp. 135-150

9

Preparing Teachers to Teach Mathematics Within a Constructivist Framework: The Importance of Listening to Children

Beatriz S. D'Ambrosio
Indiana University Purdue University Indianapolis

The goal of this paper is to raise questions that mathematics teacher educators must address as we strive to define constructivist teaching in mathematics. In the paper, I describe three forms of listening typically found in the mathematics classroom: evaluative listening, interpretive listening, and hermeneutic listening, and suggest that hermeneutic listening is the form of listening required of a constructivist teacher. The arguments are supported with examples from preservice elementary teachers who use evaluative listening in interactions with children and contrasted with an example from an experienced constructivist teacher who uses hermeneutic listening to create a model of students' mathematical understanding that guides her teaching.

Current initiatives in the preparation of teachers and in professional development experiences strive to help teachers embrace a constructivist perspective on teaching and learning. Mathematics content courses and mathematics methods courses each contribute differently to the challenge of helping teachers become constructivist in their approach to teaching. Through mathematics content courses, teachers have opportunities to challenge their views of learning as the acquisition of transmitted knowledge. Through challenging problems and group problem solving experiences, they construct understanding of mathematical content, making sense of procedures and ideas that previously were simply rules. By reflecting on those personal learning experiences, teachers' thinking about how students learn becomes the focus of mathematics methods courses. Thus,

135

mathematics content courses and methods courses work together to help teachers reflect on how their students will make sense of mathematics and how teachers can create experiences that provide children an occasion for sense-making.

In this paper I describe the listening used by constructivist teachers as they strive to help students construct knowledge of mathematics. I provide two contrasting examples of teachers interacting with children. The first is an example of a future teacher interacting with a child about a problem via e-mail conversations, and the second is an example of an experienced teacher using the same problem with her class. I conclude by speculating on the challenges of helping future teachers understand and embrace a constructivist perspective on teaching mathematics.

Teaching From a Constructivist Perspective: A Theoretical Framework

Over the years, I have been among the many individuals who believe that teaching from a constructivist perspective is unique. There are characteristics of the practice of teachers who embrace a constructivist perspective that we must come to understand in order to support the development of this perspective among future teachers. In this section, I describe what I consider unique in the practice of teachers who embrace a constructivist perspective on learning. I have chosen to focus exclusively on the act of listening to support a teacher's construction of a model of student's knowledge, which I consider the main component of constructivist teaching (Steffe & D'Ambrosio, 1995).

According to Steffe and D'Ambrosio (1995), constructivist teachers study the knowledge constructions of their students and interact with students in a learning space whose design is based, at least in part, on their working knowledge of students' understanding of concepts and ideas. This definition implies that constructivist mathematics teachers listen to learners in ways that allow them to build a model of each learner's mathematics knowledge. My recent analysis of the practice of constructivist teachers has led to new insights on characteristics of this unique practice, which I share in this paper.

Currently, I define a constructivist teacher as one who integrates the multiple voices that emerge during an instructional episode in order to shape and orchestrate the learning space for children. This teacher integrates the **voice of the discipline** and the **voices of the children**, as she defines her **inner voice** in making sense of learners' understanding.

The **voice of the discipline** includes ways of thinking, strategies, and understanding of the content that the teacher has acquired in her own learning experiences. The teacher's beliefs about the nature of mathematics shape her relationship with the discipline. From this relationship comes the teacher's own approaches towards mathematical problem solving and mathematical conversations. A constructivist teacher has internalized the nature of the discipline as a constructed body of knowledge, negotiated socially by a community of mathematicians. Understanding the sociological dimensions of this community strengthens a teacher's ability to simulate the activity of this community in the learning space she creates.

The **voices of the children** include the ways in which children make sense of an idea. The teacher elicits and draws out children's mathematical ideas by listening and watching as children engage in mathematical activity. The voices of the children are the primary sources of data for teachers who strive to make sense of children's mathematics in order to build a model of each child's understanding.

The **inner voice of the teacher** develops as the teacher integrates the voices of the children and the voice of the discipline. It includes the sense that the teacher has made of the content, the tools that she believes best illustrate the concepts, and her understanding of the nuances involved in learning the content, all of which constitute the pedagogical content knowledge of the teacher. The inner voice of the teacher includes the teacher's ability to "unpack"[1] formal mathematics in order to understand children's mathematics and to build a working model of the children's understanding.

The metaphor of unpacking is useful in describing the depth of formal mathematics understanding that is required of teachers to use mathematics in flexible and connected ways. For example, in many

---

[1] The term "unpacking" is borrowed from Deborah Ball.

mathematics classrooms, division of decimals requires learning the typical procedure of "getting rid of the decimals." Teachers explain that $156 \div 3.25$ can be solved by multiplying both numbers by 100 so that the answer to $15600 \div 325$ also solves the previous problem. Teachers often approach this explanation by representing the division as a fraction, $\dfrac{156}{3.25}$ is equivalent to $\dfrac{15600}{325}$.

The dilemma some teachers face with their mathematics knowledge lies in their ability to interpret student work on this problem such as that in Figure 1. The interpretation of this work requires a different approach to understanding the mathematics involved in division with decimal numbers. It requires what Ma (1999) suggests as a "profound understanding of fundamental mathematics" (p. 120). To interpret this work, the teacher must have different connections that will support her understanding of division. She draws on her understanding of what happens when two numbers are divided (how is dividing by 13 related to dividing by 6.5 or 3.25), what types of division problems result in the same answers, and what representations might help interpret and explain the student's work.

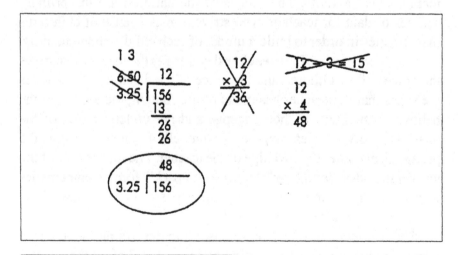

Figure 1. Student work presented in Thompson & Briars, 1989, pg. 23. (Reprinted with permission from the *Arithmetic Teacher*, © 1989 by the National Council of Teachers of Mathematics. All rights reserved.)

When pressed to explain the procedure for dividing decimals in $156 \div 3.25$, many teachers erroneously use the phrase, "what you do to one side you must do to the other," referring to 156 and 3.25 as "sides." These teachers are grasping at ideas they have committed to memory, being greatly hindered in their ability to reason about the mathematics involved in this simple problem. In spite of their fragile knowledge regarding procedures in dividing decimal numbers, teachers are successful in most mathematical tasks requiring this skill. It is only in light of interpreting students' work, which differs from traditional procedures, that teachers' inability to "unpack" the formal mathematics is revealed.

Clearly, the act of unpacking learners' mathematics requires listening to students. I interpret the term *listening* broadly to include attending to all types of "utterances" or products produced by students, such as student talk, student work, and student actions. Three types of listening typically occur in classrooms: evaluative listening, interpretive listening, and hermeneutic listening (Davis, 1996).

Evaluative listening is the type of listening used by teachers who only use the **voice of the discipline** in interpreting children's mathematical understanding. This teacher knows the rules and logic of mathematics and questions the child to search for errors in the child's thinking. I contend that evaluative listening is not sufficient to help the teacher build a model of the child's mathematics.

Interpretive listening is the type of listening used by teachers who use the **voice of the child** in interpreting the child's mathematical understanding. This teacher listens to the child and tries to put herself in the child's place, asking, "How might I be thinking about this if I had had some of the same experiences?" The teacher who is an interpretive listener attempts to make sense of what the child says or does. This teacher strives to give reason to the child.

Davis (1996) claims that interpretive listening is the listening used by a constructivist teacher. In contrast, I suggest that the constructivist teacher attempts to give reason to the child while listening to the voice of the discipline. In this struggle, the constructivist teacher often renegotiates her personal understanding of the mathematics. Hence, I propose that **hermeneutic listening** is the type of listening used by the teacher who integrates the **voice of the child, the voice of the discipline,** and **her inner voice** in order to build a model of the child's mathematics. Hermeneutic listening is a form of listening in which all parties involved in the interaction undergo some change. According to Davis (1996), "This form of listening [hermeneutic listening] is more negotiatory, engaging, and messy, involving the hearer and the heard in a shared project" (p. 53). In this sense, the teacher who listens hermeneutically is herself reconstructing her understanding of the mathematics and challenging her own ways of knowing particular mathematical ideas, based on the ideas raised during an interaction.

In summary, a constructivist teacher is one who uses hermeneutic listening to integrate the multiple voices that emerge during an instructional episode. This teacher understands the need to gain insights into the students' understanding in order to mold and shape new instructional episodes for a learning space in which students and teachers co-construct knowledge and meaning. The constructivist teacher has a passion for learning and enacts that passion as she and the children work together to make sense of mathematics in the learning space she designs.

Listening to Children

Because listening to children is an important component of constructivist teaching, I focus this section on examples of teachers listening to children. Included here are two experiences that I analyzed to gain insights into the process of listening to children. The first example describes a future teacher communicating with a child through e-mail conversations, a "virtual" field experience during a mathematics content course. The second example describes an observation of a middle-grades classroom teaching episode with a talented constructivist mathematics teacher. Through these examples I describe the act of listening that the future teacher used as she tried to understand the mathematics of the child and contrast it to that of an experienced teacher who engages in hermeneutic listening.

Analyzing E-mail Conversations

To create early opportunities for future teachers to interact with children, I inserted a virtual field component into the mathematics content course for elementary teachers[2]. My goal was to create opportunities for future teachers to engage in listening to children and to construct a model of children's mathematics. Each future teacher was assigned a partner in a sixth-grade classroom at a local school. The sixth-grade teacher and I agreed to work on some of the same problems, with the children writing their solutions to the problems for their college partner (i.e., a future teacher). The college partner responded with questions and ultimately tried to build a model of the child's mathematics. This "performance task" became a revealing source of data about the mathematics of the children and the future teachers. My goal here is to exemplify the type of listening typical of preservice teachers and to discuss the challenges involved in helping teachers to become hermeneutic listeners.

One of the problems solved by the sixth-grade children was the following:

> Three pirates were hunting for treasures and found a hefty one. They decided to split it evenly. Since it was late they thought they should sleep and split the treasure the next morning. During the night one of the pirates got up and took $\frac{1}{3}$ of the treasure and ran away. A second pirate woke up and took $\frac{1}{3}$ of what he saw and ran away. The third pirate got up in the morning and looked for the others. Trusting that they had each taken their fair share, he took what was left for himself. Did all three pirates get their fair shares? If not, which pirate got more and which got less?

---

[2] Our elementary certification spans the grades K-6.

The sixth graders had various ways of approaching this problem. Some used numbers to help them think about the problem, some drew and described pictures, and some referred to fraction amounts of $\frac{1}{3}$, $\frac{2}{3}$, and $\frac{4}{9}$. Few children used the language of operations with fractions to describe their intuitive solutions.

The sequence of conversations and the follow-up questions posed by the future teachers in many cases demonstrated their own fragile understanding of the mathematics involved in this problem and illustrated how easily they were thrown by children's solutions that diverged from their own ways of thinking about the problem. Although most future teachers were able to solve this problem themselves, often in ways similar to those of the sixth-graders, many times the solutions of the future teachers did not match that of their partners.

One future teacher received the following solution from her sixth-grade partner:

> were (*sic*) doing this problem in math class and the question is (which amount of the pirates got the largest amount of treasure)? Well the first one got the right amount because there were 3 pirates and he took $\frac{1}{3}$ of the treasure. The third [*sic*] one only took $\frac{1}{3}$ of what treasure was left so there for [*sic*] he didnt [*sic*] take the right amount. He should of (*sic*) taken $\frac{1}{2}$ of the rest of the treasure. So the 3rd one got most of the treasure. So just to refresh you [*sic*] mind the first one got the right amount, the second one got too little amount, and the third one got the greatest amount of treasure. (3/8/00)

Feeling at a loss as to how to respond or what questions to ask, the future teacher wrote:

> I need to try and understand a few things. What method
> did [sic] use to solve this problem? Can you use other
> methods? What are they? With the letter you sent me
> I see that you used subtraction. Can you use division?
> Why or why not? (3/21/00)

The comment about subtraction was not related to the child's solution to this problem, but rather referred to another problem the child had solved. Still, the tone of this message suggests that this future teacher was pushing the child to articulate a procedural solution to the problem. Not understanding this comment, the child responded:

> Well, I read your letter and yeah you can use division
> on the fair share but I cant [sic] really understand how
> to explain in writing. ...(3/22/00)

The child acknowledged the future teacher's comments and then explained the solution to a different problem. The conversation about this problem ended as the child moved to another topic.

This future teacher offered the following analysis of this exchange as well as her comments about the child's (pseudonym Angie) abilities in mathematics.

> [Angie] had the most difficult time with Math and
> sometimes would not solve the problems that she had
> been given. At the beginning of the semester she told
> me not to push her to solve the problem so I had to
> find another technique to boost her thinking. For
> example, with the fair share problem, she could
> explain the method she used to solve the problem with
> this quote: "I can use division on the fair share
> problem, but I can't really understand how to explain
> it in writing." Angie gave me the fraction of the
> amounts that was taken by the pirates in this quote:

Well the first one got the right amount because there were 3 pirates and he took $\frac{1}{3}$ of the treasure. The third [sic] one only took $\frac{1}{3}$ of what treasure was left so there for he didnt (sic) take the right amount. He should of (sic) taken $\frac{1}{2}$ of the rest of the treasure.

Angie's reasoning of how she solved this problem did not reflect the above statement in this quote: *So the 3rd one got most of the treasure. So just to refresh you [sic] mind the first one got the right amount, the second one got too little amount, and the third one got the greatest amount of treasure.*

With her explanations I was not convinced that she understood the problem. With Angie I was not surprised with her thinking, because she modeled the average math student.

There were several children who had explained that the third pirate got the most treasure because the second pirate took less than $\frac{1}{2}$ of the treasure that was left by the first pirate. Unfortunately, several of the future teachers were unable to make sense of this way of describing the partition of the treasure. The future teachers were thinking of the second pirate's share as $\frac{2}{9}$ (from $\frac{1}{3}$ of $\frac{2}{3}$). The future teachers' lack of flexibility in their thinking resulted in a view of these children's answers as "incorrect" and "lacking logic" and their mathematical abilities as average or not satisfactory.

Because this future teacher's evaluative listening was geared towards procedures, she misinterpreted the child's understanding of the problem. An evaluative listener more able to unpack the formal mathematics (i.e., one with a deeper understanding of mathematics) would likely have recognized the power of the child's representation for this problem and might have acknowledged the solution as partially correct. This listener would consider the solution incomplete, expecting the child to find the fractions $\frac{2}{9}$ and $\frac{4}{9}$ for the shares of the second and third pirates, respectively, thus matching the listener's formal mathematical solution. Evaluative listeners are listening for what they consider the "correct" answers expected of the children, in which their own levels of mathematical understanding shape what counts as a correct answer.

The interpretive listener might have recognized that the child's reference to division had been triggered by the unrelated question raised by the adult. The interpretive listener might have further speculated whether the child thought formally about division in her original explanation, or whether there was a different model serving as a referent for the child's solution. This teacher would be likely to hear the voice of the child and attempt to make sense of that voice.

The future teacher exemplifies what we might anticipate in an evaluative listening episode. In contrast, the following example provides an opportunity to discuss the hermeneutic listening of a constructivist teacher.

Observing a Teaching Episode

The following discussion, with elaborations of my field notes, relates to a teaching episode in a middle-grades mathematics classroom. A multi-age classroom of middle-school children in sixth through eighth grade worked on the treasure problem. The teacher posed the problem and suggested that the children discuss solutions in their small groups, preparing a presentation for the rest of the class. The teacher expected the problem to be a warm-up before engaging in other activities involving fractions. Unexpectedly, this problem took the full class period to unravel.

Having listened to the children by observing them working in small groups and articulating their solutions to each other, the teacher determined the order in which solutions would evolve during the large group presentation. The most mathematically sophisticated presentations were left for the end of the session, while the most incomplete solutions served as the springboard for the large group discussions. This procedure served to avoid premature closure for students who had not yet constructed a full explanation of the solution. The teacher's purposeful sequencing of the solutions resulted from her use of hermeneutic listening. Rather than describe the entire sequence of presentations, I analyze the teacher's analysis of one child's thinking based on his work and presentation to illustrate further her use of hermeneutic listening.

Midway into the large group discussion, Eric willingly accepted the invitation to be the next to share.

> Eric started his presentation by drawing nine tiles on the overhead projector.
>
> Eric: "I asked myself, what can make 3, 3, 3? That's 9."
>
> Eric: "Pirate number one got $\frac{3}{9}$, pirate number two gets $\frac{3}{6}$, he's supposed to get three. Pirate number three gets $\frac{3}{3}$. You have to find a number that equals all of their treasure."
>
> T: "So Eric, which pirate got more?"
>
> Eric: "They each got the same. They each got 3."

The teacher was intrigued by this student's work. Her intrigue stemmed from many perspectives, including her own understanding of the mathematics of this problem, her knowledge of students' difficulties in working with fractions, her repertoire of solutions she had witnessed in earlier problems, and her evolving model of this child's mathematical understanding. She and Eric and the other members of the class were partners in searching for understanding and co-constructing meaning and reasoning to support Eric's solution.

The collection of student work and conversations during the presentation served as data for the teacher to analyze as she herself sought new understandings for this problem.

Although there is not enough space to articulate all of the ideas that this teacher constructed from her reflection on this episode, I summarize a few insights that became a springboard for further investigation of students' understanding. Eric's thinking was shaped by his firm belief that the portions should be fair, so he sought to determine fair shares of the treasure. The most important struggle revealed in this child's work was his attempt to use the language of mathematics to discuss each pirate's share. With each new piece of the story, a new whole was defined. Eric articulated each pirate's share as a fraction of a new whole. Eric's explanation was compelling to his colleagues, which intrigued the teacher even further. Were other students trying to use the language of mathematics and struggling with the apparent whole in each piece of the story? Why had Eric's solution resonated with so many of the students? Her questions addressed the mathematical nuances involved in this simple task. Why would such a task lead Eric to respond in the way he did?

The hermeneutic listener views her own learning as a process of inquiry. This teacher encouraged the children to explain their answers and their reasoning to others, in the hope that this conversation might generate in the children the same level of curiosity and intrigue that it had generated for her. She was sensitive to the fact that there would be other questions emerging in the children's minds (ultimately more interesting to them and to her) as they explored this seemingly simple mathematical question.

The hermeneutic listener engaged with the children in a joint inquiry project, based on this problem. The teacher negotiated multiple voices, including the voice of the discipline, the voice of the child, and her inner voice to gain new insights and construct new understandings for this problem and the children's mathematics.

In contrast, an **evaluative listener** would consider Eric's response incorrect and plan strategies to change his response to the "correct" one. An **interpretive listener** would try to make sense of his response from the perspective of the child to plan how to help the child. She might ask herself, "what led the child to this response? What made the child believe that this would be the answer?" Teachers who are interpretive listeners trust their students' thinking and work hard to make sense of students' responses. Although one might say that such a teacher is truly constructivist, I claim that such a teacher has an emerging understanding of the constructivist perspective.

In this article I have purposefully not included much of the dialogue between the experienced teacher and her students. I hope to convey to the reader that the different forms of listening are not a result of the questions asked by the teacher, but rather a result of how the teacher attends to the children's responses, regardless of the types of tasks or questions posed or the dynamics of the class discussions. It is conceivable that the same questions could be used by an evaluative listener, an interpretive listener, or a hermeneutic listener. The dialogue that results could be almost identical. Yet, the different listeners would attend to the voice of the children in three very different ways.

As the work of Schifter and Fosnot (1993) suggests, teachers become more able to enact teaching within a constructivist perspective as they shift their focus, when reflecting on teaching, away from their own teaching behaviors and replace it with a focus on the children's learning and their needs. Schifter and Fosnot (1993) describe this shift in focus (from a focus on teacher behavior to a focus on children's learning) as stages of development. The transition from evaluative listening, to interpretive listening, to hermeneutic listening seems to parallel the shifts described by Schifter and Fosnot (1993). The hermeneutic listener is the one who is best able to integrate the voice of the discipline, the voices of the children, and her own inner voice in designing a learning space that creates opportunities for children to construct meaning and understanding of mathematical ideas, thus characterizing one of the primary activities of a constructivist teacher.

Conclusions

The involvement of preservice elementary teachers with children through e-mail conversations helped preservice teachers "tune into" children's understandings. The assignment was grounded in the belief that the task of making sense of children's understanding can promote teacher's content knowledge and enhance teacher's understanding of the importance of listening to children. Preliminary evidence suggests that the preservice teachers who were involved in this e-mail project were more thorough than those who had not had this opportunity in completing assignments in their mathematics methods courses that required them to interview children or analyze children's mathematical work; however, they primarily used evaluative listening in the construction of a model of children's mathematical understanding.

Throughout this work I have become increasingly troubled by the question, "what does it take to ensure that future teachers will be able to interpret children's work with flexibility and an adequate depth of understanding?" There are few examples of future teachers who successfully use interpretive listening, or seem to be developing an emerging understanding of constructivist teaching. We have no examples of future teachers who successfully use hermeneutic listening and demonstrate a solid understanding of constructivist teaching. Is it unrealistic to expect that a teacher education program can successfully prepare constructivist teachers? I would prefer to think that we have not yet discovered what it takes to develop a constructivist teacher. As a community focused on the preparation of teachers, it is not clear to me that we have designed the learning space for future teachers in a way that facilitates their growth as constructivist teachers. As a result, it is not clear that they will have a disposition towards the use of hermeneutic listening as they create learning spaces that facilitate children's construction of mathematical knowledge. With this discussion I hope to challenge the community of mathematics teacher educators to explore, in collaboration with all of our colleagues in teacher education programs, the process of becoming a constructivist teacher.

References

Davis, B. (1996). *Teaching mathematics: Toward a sound alternative.* New York: Garland.

Ma, L. (1999). *Knowing and teaching elementary mathematics: Teachers' understanding of fundamental mathematics in China and the United States.* Mahwah, NJ: Lawrence Erlbaum Associates.

Schifter, D., & Fosnot, C. T. (1993). *Reconstructing mathematics education: Stories of teachers meeting the challenge of reform.* New York: Teachers College Press.

Steffe, L., & D'Ambrosio, B. (1995). Toward a working model of constructivist teaching: A reaction to Simon. *Journal for Research in Mathematics Education, 26,* 146-159.

Thompson, A., & Briars, D. (1989). Assessing students' learning to inform teaching: The message in NCTM's evaluation standards. *Arithmetic Teacher, 37(4),* 22-26.

Beatriz S. D'Ambrosio, Associate Professor of Mathematics Education at Indiana University Purdue University Indianapolis, has been on the faculty since 1994. Previously, she was a faculty member at the University of Campinas (Brazil), University of Delaware, and University of Georgia. Her main research interest is the study of constructivist teaching. Linked to this work is the study of the professional development of teachers through preservice and inservice education. Her work has been published in journals nationally and internationally, including in the *Journal for Research in Mathematics Education, Educational Studies in Mathematics, School Science and Mathematics,* and in several book chapters. [bdambro@iupui.edu]

Taylor, A. R. and O'Donnell, B. D.
AMTE Monograph 1
The Work of Mathematics Teacher Educators
©2004, pp. 151-167

10

Revealing Current Practice Through Audio-Analysis Releases the Power of Reflection to Improve Practice

Ann R. Taylor
Barbara D. O'Donnell
Southern Illinois University Edwardsville

If teachers' practice is partially hidden by their current beliefs and they reflect on their perceived practice, then reflection will not likely move them forward. The audio analysis assignment was developed to enable preservice and inservice teachers to examine their practice or "cultural script" through transcribing their interaction with a student and analyzing the results. Teachers ask hard questions of themselves about the effectiveness of their mathematical practice as they confront their own words on tape. Reflecting on actual practice releases teachers' ability to learn from and improve their instruction. Examples of teachers' analyses and rubrics for scoring the assignment are provided.

Professional development needs to help teachers change their practice based on research on learning and teaching mathematics. Much has been written about using reflective thinking (Schön, 1983) and metacognitive processes (Bransford, Brown, & Cocking, 1999) to enhance learning. But what if teachers' perceptions of their teaching are inaccurate? If teachers are reflecting on their perceived practice, then reflection will not likely move them forward. Stigler and Hiebert (1999) recognized the persistence of "cultural scripts," noting that even when U.S. teachers thought they were using reform practices, this was rarely the case.

In the audio-analysis assignment described in this paper, teachers examine their practice or "cultural script" through transcribing an audiotape of their teaching. The assignment provides a powerful opportunity for learning set in the context of teachers' own practice by a) challenging teachers to reflect on their current practices based on an accurate understanding of their own mathematics teaching, b) deepening teachers' skills of inquiry and critique, and c) enabling teachers to set their own agenda for professional growth, in light of new understanding of their own practices.

The audio assignment is related to, but significantly different from, common preservice assignments of child study, child assessment, or diagnostic interviews (Huinker, 1993; Moyer & Milewicz, 2002). This audio-analysis assignment requires teachers to analyze their mathematical "talk," their skill in formulating questions, and their responses to perceived student thinking and understanding.

Assignment Description

After obtaining permission to audiotape, each preservice or inservice teacher tapes a twenty-minute exchange with one or two students. As preparation, we engage in a class discussion of "Questioning Your Way to the *Standards*" (Mewborn & Huberty, 1999) to learn about effective and ineffective questions typically asked by teachers; in addition, class members are provided with a list of possible questions for use in the assignment. We watch the TIMSS *Eighth Grade Mathematics Video Lessons from the United States, Japan, and Germany* (U. S. Department of Education, 1999), which contrasts teaching as lecture-demonstration with teaching through problem solving and critical questioning.

For the audiotaping, teachers may choose to review an assignment with a student, work on concepts that challenge the student, or even teach the student a new concept or skill. Preservice teachers seek the advice of their classroom teacher when selecting the subject matter and interviewee.

The teacher listens to the tape and chooses the most revealing five-minute segment to transcribe. Once the verbal exchange is in written form, teachers add between-the-lines analysis to the transcript, including critical comments and suggestions for improving the dialogue. Negative critical comments might include statements about leading the student to an answer, use of improper mathematical language, faulty content knowledge, explanations that were ineffective, praise of correct answers, or telling students procedures instead of asking questions that support them to construct their own understanding of mathematical concepts.

The completed assignment consists of the following parts:

- an introductory paragraph with an overview of the school, the students, and the format and text used for typical mathematics lessons;
- the transcript and the between-the-lines analysis;
- a summary indicating what they learned about themselves and their students;
- an indication in either the summative remarks or the between-the-lines analysis about how the research from the textbook or course readings informs their practice.

To help teachers understand between-the-lines analysis, we demonstrate the process with a short example and later facilitate a workshop in which teachers critique peers' rough drafts. We have used two formats for this assignment. In the first format, teachers complete two separate audio analyses but there is no peer critique. Many teachers, especially preservice teachers, are unable to analyze the interviews in-depth on their first attempt. They rely on our critical comments to draw their attention to weaknesses in their interaction. Later in the semester, a second audio analysis is conducted. Analyses are greatly improved; only the higher of the two grades from the two audio analyses contributes to the final course grade. In the second format, teachers complete only one audio-analysis assignment and rely heavily on peer critiques.

The first format requires a greater time commitment on the part of the instructors who are the principal reviewers of the teachers' work. The second format results in equally good analyses because teachers assume the task of critiquing. We choose the format appropriate for a particular class based on other assignments and course goals; the assignment is usually worth 30-40% of the final course grade.

We evaluate the audio-analysis assignment with a detailed rubric that assesses three specific traits. The detailed language provides teachers with guidance on completing the assignment (see Appendix). In the first trait, knowledge of learners and the process of learning mathematics, we examine teachers' ability to understand what their students are thinking. In the second trait, directing discourse, we analyze the dialogue to determine if teachers ask questions that require students to reason and justify answers or if their questions lead the student to a one-word answer. If teachers only complete one audio analysis, we interpret this section according to their ability to identify and offer specific suggestions on correcting issues. The third trait, analysis of mathematical discourse, is the most important and weighted most heavily in the grading. The between-the-lines analysis demonstrates whether teachers can detect problems in their own discourse, suggest specific rewording, and use their own selection of research readings to inform their analysis. The following excerpt is from the analysis of a preservice teacher, Karen, who worked with a student on using divisibility rules to find the factors of a number. The excerpt illustrates how we comment on the analysis and use the rubric in the Appendix; italics indicate the teacher's between-the-lines analysis and faculty comments are in bold type.

Student: The sum of the digits is divisible by 3.
Teacher: Okay, so what do you think you would do then with that number ... 80?
Instead of asking the student what she would do with the number, I should have let her figure it out on her own. **Another option is to rephrase the question. How could you do this without assuming specific student knowledge?**
Student: Just take the 8.
I assumed that she added the 8 and the zero. I should have asked her how she arrived at the answer. **I agree. You cannot assume that she is using the divisibility rule. This would also be an appropriate time to comment on procedural or conceptual knowledge.**

Teacher: Okay.
Student: And see if it is divisible by 3.
Teacher: Okay. So is 8 divisible by 3?

This question does not seem appropriate for this student. I should have rephrased this question. **How would you rephrase your question? What should you have done differently? You seem to be using "okay" to validate student answers; do you mean to do that?**

When we examine Karen's transcript according to the three traits, we realize that she is attempting to discover what the student understands about divisibility rules; however, Karen's questions are not making the student explain her thinking. Without revealing evidence, Karen's statements about the student's knowledge may be unfounded. Karen also neglects to reference course readings on conceptual and procedural knowledge. Although she does well to acknowledge discourse problems, she does not always suggest what she should have done to improve the dialogue. When assessing this small excerpt using the three traits, we determine that Karen's attempts are "sprouting" in all three areas.

Examination of Inservice Teachers' Audio-Analysis Assignments

Through a qualitative analysis of the reports written by 27 inservice teachers enrolled in a graduate course, we found that categories of reflection and self-awareness were central to the value of this audio analysis. Thus, we categorized teachers into three general groups using reflectiveness and self-awareness as separate axes: Group 1 – High Reflectiveness/High Self-Awareness; Group II – High Reflectiveness/Low Self-Awareness; and Group III – Low Reflectiveness/Low Self-Awareness. Nearly all teachers were reflective about their teaching. Some reflective teachers were already aware of their practice. However, the majority of reflective teachers were surprised and shocked by crucial aspects of their mathematical practices with children.

Group 1 - High Reflectiveness/High Self-Awareness

About 10-20% of this group of teachers have well-developed instructional practices and have thoroughly thought about why they do what they do. So, their audio analysis is less a discovery and more an opportunity to share and refine sophisticated thinking, justify their decisions with carefully articulated reasoning, and push themselves to extend their discussion with children even further. In this example, Barbara is working with kindergarteners on the inverse operations of addition and subtraction, spending more time than usual on this topic.

> **Levi:**   Six take away three is three.
> **Teacher:** Okay. How do you know that? Several others have said that. How do you know that six take away three is three?
> **Levi:**   Because. Because we used it in our head.
> **Jordan:**   OH! OH! Because three plus three is six!!
> **Teacher:** Oohhh! Now there's an interesting problem. I don't think anybody thought of that this morning.

This was a big revelation and a way to explain a mathematical connection. I am always intrigued by things students see and think about that aren't in my 'lesson plan.' This questioning allows students to make observations and/or discoveries that often times takes a lesson in a different, but valid, direction. It can leave a teacher feeling vulnerable and unprepared. A "Tell me more" statement may have led the students more directly to what they would be able to discover and explain to the others. Interestingly, no journal writings reflected this idea that was developing. Only out of our discussion did this concept develop. According to Ma, manipulatives are only useful if discussion follows. "In contrast to the U. S. teachers the Chinese teachers said they would have a class discussion following the use of manipulatives in which students would report, display, explain, and argue for their solutions" (Ma, 1999, p. 26). David picks up with a further explanation.

> **David:** HEY! But you can do it backwards!
>
> **Jordan:** Yeah... you can go backwards.
>
> **Teacher:** Like so...
>
> **David:** Like if the two three's are in the front. It would be three plus three equals six. But if you do it backwards you put the minus there. It would be...
>
> **Teacher:** So if you go backwards you can take six...
>
> **David:** Away
>
> **Teacher:** Take away three - one of those you had on the other side? And you end up with something else you had on the other side?

This would have been a good place to ask, "Can you convince the rest of us that that makes sense?" It seems we were all (especially the teacher) having a bit of trouble expressing what we knew!

Barbara seems to dance through her dialogue with a sense of direction, joy, and curiosity. Her reflection is so developed that we termed her group of highly reflective teachers "ponderers." These ponderers do more than identify, think about, and justify their actions or words. They extend their reflection by opening their interactions to a second round of reflection and questions. They often express uncertainty about what they have done and thought, even when apparently successful, because they recognize the possibilities that exist in any moment of teaching.

Group II – High Reflectiveness/Low Self-Awareness

The majority, about 60-80% of this group of teachers, is not very aware of what they do. However, when provided with an opportunity to hear themselves during instruction, they are quick to critique the inadequacy of their practice using research literature and ideas from the course.

In this sample dialogue, Diana is working with a 4th grader on multi-digit multiplication.

At this point I am assuming that Katie has been instrumentally taught. I know after teaching in the system, in several states, the issues that Skemp (1978) brings out are very common with our system of teaching. The ideas that instrumental mathematics is usually easier to understand, rewards are more immediate and more apparent, and one can get the right answer more quickly are brought up when there is any question about changing our styles of teaching.

Teacher: Well here, write 56 out to the side, All right, tell me about 56.

I know I fell into the idea of not wanting to frustrate her, afraid that she might give up. Katie is a very intelligent girl with a high IQ, but has a hard time accepting not being able to come up with an instant answer.

Katie: It's got two digits.
Teacher: It's got two digits. Anything else about it? Don't worry about this. (Pointing to the original problem) Tell me about 56. (Long pause) Anything, anything at all? Think of your place value.
Katie: It's in the tens.
Teacher: What's in the tens?
Katie: 56... 50
Teacher: 50 is. The-5-is-in-the-tens.

I really felt myself leading Katie down a certain path at this point. I was wanting her to do more thinking on her own, but didn't feel I was accomplishing it at this point.

Katie: The 6 is in the ones.
Teacher: OK. Now do you know why you split that up over there?
Katie: Because the ones go down here and the tens go up over here.
Teacher: Because that's the what column?

Katie: The tens column.
Teacher: Good. Go on from there.

[Summary at end of analysis] *After doing this activity, I really became aware of how I have been a procedural/ instrumental teacher much more than I thought. It takes more conscious effort to become the conceptual/ relational teacher that I want to be. I have to learn to be more patient and give time for the students to think. I tend to want to guide them along a narrow path, not giving them the chance to stray off of the path; afraid they may get lost along the way. I am also starting to realize how uncomfortable our students are with the open-ended questions. The students are so programmed to just do the problems very procedurally and not even think about the meaning of the process. Even students with high IQ and gifted students want an immediate response if their answer is right or wrong. This puts such a squelch on their thinking process.*

Another fourth-grade teacher expressed shock after she transcribed and analyzed her teaching of double-digit multiplication.

The bell rung and the thirty minutes of torture was over for them. I can't believe how much I talked and controlled the pulse of the whole conversation and lesson. I did not ask enough questions of any kind. I didn't ask for any clarification or further explanation at any time. ... My students could only give very short answers to my questions because that was all they were allowed to do. I was the one doing most of the mathematical work and controlled everything.

A few teachers in this group were also "ponderers," offering extended reflections about their thinking and actions, even though their awareness was initially low.

Group III – Low Reflectiveness/Low Self-Awareness

The third group, about 10-20% of these teachers, was unable to consider the implications and possibilities of hearing their own practice. They may recognize a weakness, but they justify it rather than express concern for its impact on students. These teachers tend to focus on superficial aspects of mathematical practice, such as the prevalence of informal language or unfinished sentences. They offer few alternatives for what they might have said.

Examination of Preservice Teachers' Audio-Analysis Assignments

Preservice teachers enter mathematics methods courses with definite ideas about teaching mathematics. Hence, we give this assignment at the beginning of the semester to ensure its impact throughout the course. In Spring 2003, preservice teachers ranked the audio analysis as the most effective assignment, with 34 out of 56 naming it the pivotal assignment of the course. These preservice teachers fell into the same broad categories as the practicing teachers, except there were fewer in the High Reflectiveness/High Self-Awareness group.

Primary themes emerging from preservice teachers' writing included self-awareness of the ineffectiveness of their current knowledge and practice, a shift to focusing on student learning, and a perceived need to make changes. The first theme emerging from the data was self-awareness. Many preservice teachers espouse reform ideas, but do not see the depth and complexity involved in teaching mathematics. However, when confronted with undeniable evidence of unproductive teaching behavior, preservice teachers became aware on many levels. They developed a personal awareness of their limited use of mathematical language. Many became confused when they could not describe or effectively form questions to help their students. They did not, however, ascribe this problem to lack of mathematical knowledge. Preservice teachers also became aware of ineffective teaching practices. As one preservice teacher stated, "I found that at times I was confusing and leading students to answers. After completing this assignment, I am much more aware of some of my weaknesses and areas I need to improve on."

We also saw a shift in preservice teachers' instructional focus from solely targeting their teaching and questioning skills to beginning to think about student learning. As one preservice teacher so aptly put it, "It [the audio analysis assignment] also made me think of how I want my students to learn math. Do I simply want them [students] practicing problems that I give examples for or do I want to challenge their intellectual abilities by posing worthwhile problems that they have to figure out how to answer?" Another stated, "This was a good assignment to see the kinds of questions one asked the students. I realized I asked a lot of leading questions that allowed the students to answer the questions very easily."

The increased awareness and shift to consider student learning led preservice teachers to analyze changes they need to make: "Using this audio analysis experience, I began to modify my questioning and thinking in the classroom. Whenever teaching lessons in mathematics, I would pay close attention to how I asked questions and would think about ways to force the students into conceptual thinking."

Reflecting on the Assignment

After reflecting on the continuum of experiences in this assignment, we share the following observations. The systematic, self-analysis of the audio transcript is powerful. Teachers recognize and make changes in their "cultural script" because they are confronted by indisputable problems in their teaching. The process of transcribing their teaching dialogue slows down what Lewis and Tsuchida have called the "swiftly flowing river" of teaching (1998). Some teachers have listened several times to the tape when transcribing to hear what they were saying, because it was so different from what they had expected.

Adding between-the-lines analysis deepens teachers' reflection and helps them attend systematically to the impact of their actions on students' mathematical thinking. They consider the differences between students' knowing conceptually and procedurally. Research surrounding student learning becomes relevant and enlightening. Frequently, teachers provide reasons why their imagined interactions would benefit students and cite language from mathematics literature to support their statements.

Teachers completing a second audio analysis showed evidence of beginning to make a shift in their instructional practice. Lisa, a "ponderer" from the Highly Reflective/Low Self-Awareness Group II, works with her kindergarten children in the following dialogue exchange:

> **Teacher:** How many beans do you suppose are in the cup?
> **Student:** Eight?
> **Teacher:** What makes you think there are eight beans in there?

Instead of accepting the answer as eight, I heard the question in his voice, so I thought I would pursue it further to see if he got the connection as to why there would be eight in the cup.

> **Student:** Because when we explored the other numbers there were that many.

It would have been easy for me to accept the answer that he gave, because I did know what he meant. However, since my first audio-analysis I've done a lot of reflection on how I phrase things and sometimes "set the children up" to just fill in the blanks, so I'm trying to be ready when a child gives me part of the answer I want, but not the entire answer. Hoping I can get them to be more complete in their answers and be able to expand a little further with their language.

> **Teacher:** There were eight?
> **Student:** No, I mean when we did six there was six and when we did five there was five and now we're doing number eight.
> **Teacher:** Oh, I see what you mean now. Go ahead and spill the beans and tell me about it.
> **Student:** I've got three reds and five whites.
> **Teacher:** Okay, how will you record that on your

paper?

Before my first analysis I don't think I would have even asked how, I would have just told him to record it. It has really dawned on me how important it is to explain and verbalize more. I now realize that the deep comprehension of what one is doing comes from talking about what one is doing.

Conclusion

Classroom teachers of mathematics from preservice to veterans need opportunities to strengthen their disposition and skills to study and improve their teaching. Feiman-Nemser (2001) argues that, "to continue learning in and from teaching, teachers must be able to ask hard questions of themselves and their colleagues, to try something out and study what happens, to seek evidence of student learning, and explore alternative perspectives" (p. 1040).

The assignment we have described develops teachers' abilities to ask hard questions of themselves about the effectiveness of their own mathematical practice as they are confronted with the indisputable reality of their own words on tape. Teachers explore alternatives and extend their repertoire of possible interactions in their between-the-lines analysis.

The assignment supports preservice teachers' learning by challenging their assumptions that they are experts. Teaching mathematics is not the simple task they first thought, and candidates recognize how easily they might unintentionally teach students using rote procedures. Assigning the audio analysis early in the course is critical because teacher candidates then approach other assignments with a conviction that they need to learn how to help their students construct an understanding of mathematical concepts.

We believe the audio analysis contributes to the learning of practicing teachers by a) increasing teachers' self-awareness of their current practice, b) helping teachers extend and deepen their reflection by analyzing their practice in relation to concepts of reform literature, and c) enabling teachers to set their own agenda to change practice based on their new understanding and adapted to their own context.

References

Bransford, J., Brown, A. L., & Cocking, R. (Eds.). (1999). *How people learn: Brain, mind, experience, and school*. Washington, DC: National Academy Press.

Feiman-Nemser, S. (2001). From preparation to practice: Designing a continuum to strengthen and sustain teaching. *Teachers College Record, 103*(6), 1013-1055.

Huinker, D. M. (1993). Interviews: A window to students' conceptual knowledge of the operations. In N. L. Webb (Ed.), *Assessment in the mathematics classroom* (pp. 80-86). Reston, VA: National Council of Teachers of Mathematics.

Lewis, C., & Tsuchida, I. (1998). A lesson is like a swiftly flowing river: How research lessons improve Japanese education. *American Educator,* (Winter), 14-17, 50-52.

Ma, L. (1999). *Knowing and teaching elementary mathematics: Teachers' understanding of fundamental mathematics in China and the United States*. Mahwah, NJ: Lawrence Erlbaum.

Mewborn, D., & Huberty, P. (1999). Questioning your way to the *Standards. Teaching Children Mathematics, 6*(4), 226-246.

Moyer, P., & Milewicz, E. (2002). Learning to question: Categories of questioning used by preservice teachers during diagnostic mathematics interviews. *Journal of Mathematics Teacher Education, 5*(4), 293-315.

Schön, D. A. (1983). *The reflective practitioner: How professionals think in action*. New York: Basic Books.

Skemp, R. R. (1978). Relational understanding and instrumental understanding. *Arithmetic Teacher, 26*(3), 9-15.

Stigler, J., & Hiebert, J. (1999). *The teaching gap: Best ideas from the world's teachers for improving education in the classroom*. New York: Free Press.

U. S. Department of Education. (1999). *Eighth grade mathematics lessons: United States, Japan, and Germany*. Washington, DC: National Center for Education Statistics.

Ann R. Taylor, Associate Professor in the Department of Curriculum and Instruction, Southern Illinois University Edwardsville, teaches undergraduate and graduate courses in mathematics education and action research. Prior to earning a doctorate from Washington University, St. Louis, she was a middle-school and high-school teacher in England. Her research interests focus on mathematics methods courses, language use in mathematics teaching, and teacher professional development. [ataylor@siue.edu]

Barbara D. O'Donnell, Assistant Professor in the Department of Curriculum and Instruction, Southern Illinois University Edwardsville, teaches undergraduate and graduate courses in mathematics education and supervises teacher candidates' field experiences. Prior to earning a doctorate from the University of North Dakota, she was an elementary and middle-school mathematics teacher. Her research interests are reform in mathematics methods teaching and related clinical experience issues. [bodonne@siue.edu]

Appendix. Rubric for Audio-Tape Analysis[1]

Trait: Knowledge of learners and process of learning math (10%)

Flourishing	Sprouting	Germinating
Central purpose of exchange seems to be to understand the mathematical thinking of the student. Teacher demonstrates knowledge that children learn mathematics by making sense through reasoning, not by being told by the teacher.	Teacher values mathematical thinking sometimes, but is not able to pursue this. Some evidence that teacher may believe children learn by making sense, but practice does not always match this belief.	Student's own reasoning and language are not important. Teacher's agenda dominates, so there is little student learning. Teacher shows no awareness that children learn through making sense. There is evidence that teacher believes children learn by being told.

Trait: Directing Mathematical Discourse (20%)

Flourishing	Sprouting	Germinating
Main interaction is through asking questions. Questions require students to reason and justify their responses. Language does not reinforce correct answers or incorrect ones. Teacher attempts to help student to think for himself or herself.	Teacher asks some questions, but also tells students. Or, many questions are simple, requiring one-word answers. Teacher mostly decides what is correct and incorrect; students do not clarify and justify their own thinking. Teacher mostly praises correct answers and gives hints so student does not have to think.	Language gives directions, or tells students what or how to think. Purpose is to go through mathematical procedures. Teacher always praises correct answers and intervenes immediately to fix incorrect answers. No evidence that teacher expects students to think and reason.

---

[1] Assignment accounts for 30-40% of final course grade.

Trait: Analysis of the Mathematical Discourse (70%)

Flourishing	Sprouting	Germinating
Teacher demonstrates insightful awareness of teacher language and its significance to successful learning and the ability or desire to recognize productive and non-productive discourse. Teacher suggests possible re-wording of interactions and supports from literature.	Teacher shows understanding of significance of teacher's language. Teacher recognizes need to make different responses, but does not offer rewording. Or, rewording still does not seem likely to require students to reason for themselves. There are short or inappropriate references to literature.	Teacher shows little evidence of understanding the significance of the teacher's role or language in developing and shaping the discourse. Teacher seems unaware of opportunities to explore ideas in depth. There is little or no recognition of the impact of this on student's learning. There are no references to literature.

Moyer-Packenham, P. S.
AMTE Monograph 1
The Work of Mathematics Teacher Educators
©2004, pp. 169-188

11

The Interview Assignment: Evaluating a Teacher Candidate's Knowledge of Mathematics Content, Questioning, and Assessment

Patricia S. Moyer-Packenham
George Mason University

In addition to understanding children's mathematical thinking, interviews can be a useful tool for gathering information about the interviewers. This paper describes the use of clinical mathematics interviews, conducted with children by elementary education teacher candidates, as a way to assess teacher candidates in a mathematics methods course. It outlines how teacher candidates plan the interview, develop appropriate tasks and questions, and evaluate and synthesize the data in a final report. The paper provides a grading rubric and suggestions for using the assignment as an assessment tool in an elementary mathematics methods course.

Clinical interviews are frequently used to gain an understanding of children's thinking in mathematics (Buschman, 2001; Fennema & Carpenter, 1996; Fennema, Franke, Carpenter, & Carey, 1993). In addition, interviews are a useful tool for gathering information about the interviewers. Important insights can be gained by examining the tasks and questions interviewers prepare before the interview, the types of questions interviewers use as follow-up to students' responses during the interview, and the interviewers' analysis following the interview. Previous research on this topic has shown that teacher candidates who use interviews for assessment view interviewing and questioning as alternative assessment strategies for use in their mathematics teaching (Moyer & Moody, 1998). Practicing and analyzing questioning skills allow teacher candidates to reflect on the questions they use during mathematics interactions (Moyer & Milewicz, 2002).

This paper describes the use of clinical mathematics interviews to assess elementary teacher candidates in a mathematics methods course. In particular, the assignment focuses on assessing teacher candidates' content and pedagogical content knowledge in mathematics through their application of that knowledge in the design and analysis of a mathematics interview with a child. The design of the interview, the assessment of the child's knowledge following the interview, and the instructional plan created for the child based on the interview provide important information about teacher candidates' understanding of mathematics content and pedagogy.

Over the past six years, I developed and revised this assignment and grading rubric, and it is currently used as a performance-based assessment in an elementary mathematics methods course that includes field experiences. Teacher candidates design and conduct these interviews following two months of readings, discussion, and interactive class sessions that focus on mathematics content for the elementary grades and instructional tools and strategies that can be used to engage children with that content. This assignment builds on previously published models of alternative assessment for teacher candidates (Chappell & Thompson, 1994) and represents a step forward by including a detailed outline of the requirements and a rubric to assess their work.

Developing the Interview Plan

The assignment involves several important tasks that develop skills transferable to future teaching activities for elementary education majors, including: (1) designing questions and tasks based on children's developmental levels; (2) questioning children about mathematics for deeper understanding; (3) recording student data; (4) evaluating a child's mathematical knowledge; and (5) designing instructional plans based on children's performance. A variety of sources can be used to provide teacher candidates with information on developing clinical interviews with children (Cross & Hynes, 1994; Labinowicz, 1987; Long & Ben-Hur, 1991; Peck, Jencks, & Connell, 1989). The textbook that is currently used in this course is *Helping Children Learn Mathematics*, 7th edition (Reys, Lindquist, Lambdin,

Smith, & Suydam, 2004). These sources provide prospective teachers with background information on interview processes, selecting tasks and questions, and recording and analyzing data. The information provided to prospective teachers on designing the plan is shown in Figure 1.

The Child
Describe the child you plan to interview. Include information you gather about the child (grade level, age, gender, race, and academic ability level). What do you know about the child's level of understanding about the mathematics concept before the interview?

The Mathematics Concept
Select one specific mathematics concept to assess during the interview. Examples of concepts might include patterns, sorting, addition of whole numbers, division of fractions, finding averages, percent, geometric shapes, or length measurement. Tell why this concept is appropriate for this child at this particular grade level.

Different Forms of Representation
During the interview, assess the child using three different forms of representation. Identify the three different forms of representation you will use during the interview with at least one example in each form. Concrete representations include manipulatives, measuring tools, or other objects the child can manipulate during the interview. Pictorial representations include drawings, diagrams, charts, or graphs that are drawn by the child or are provided for the child to read and interpret. Symbolic representations include numbers or letters the child writes or interprets to demonstrate understanding of a task.

Tasks & Questions
Design tasks and questions that use three different forms of representation (concrete, pictorial, abstract symbols) to diagnose the child's understanding of ONE basic concept. Go beyond the basic level of determining the child's factual knowledge of the concept by asking questions that determine how much the child understands about the concept. For example, suppose you are assessing the concept of ADDITION. (1) Create several tasks where the child uses concrete manipulatives to demonstrate her understanding of addition; ask questions about the child's understanding of the addition tasks with manipulatives. (2) Create several tasks where the child is asked to create or interpret drawings to demonstrate her understanding of addition; ask questions about the child's understanding of these tasks with pictorial models; (3) Create several tasks where the child uses abstract symbols (and letters) to demonstrate her understanding of addition; ask questions about the child's understanding of these addition tasks using the symbols.

Figure 1. Student assessment interview plan guidelines

Gathering Information About the Child

To begin the assignment, prospective teachers select a child for the interview from their field experience placement in a local elementary school. They gather information about the child that includes the child's grade level, age, gender, race, academic ability level, and level of understanding about the mathematics concept that will be a part of the interview. They may also gather information about the child's performance in other academic areas or social and behavioral descriptors that provide a context for the information gathered. Prior to the interview with the child, teacher candidates obtain permission from the child's parents to audiotape and photograph the interview.

Selecting a Mathematics Concept
and Ways to Represent That Concept

To plan the interview, prospective teachers select one specific mathematics concept to assess during the interview with the child. For example, they may choose to interview a kindergarten child about knowledge of repeating patterns, or they may choose to interview a 4th-grader about knowledge of adding fractions with unlike denominators. It is important that the interview focuses on only one concept and that the interviewer examines the child's knowledge of that concept in depth. Prospective teachers sometimes find it challenging to select only one concept and to examine the concept in detail, perhaps reflecting a lack of experience in learning mathematics concepts beyond the surface level. For the instructor, the interview can provide information on the teacher candidates' depth of knowledge on the particular mathematics topic selected.

Prospective teachers are required to identify different forms of representation they will use to present and assess the concept and to provide examples of the representations. Their plans must include a task using at least one representation from each of the three different forms commonly used in school mathematics: (1) concrete/physical, such as manipulatives, measuring tools, or other objects children can manipulate; (2) pictorial/visual, such as drawings, diagrams, charts, or graphs; and (3) abstract/symbolic, such as numbers, letters, or operation signs. The teacher candidate's ability to select appropriate representations of the concept provides the instructor with insights about his/her level of understanding. For example, can the teacher candidate design tasks to represent the concept in a variety of ways and make connections among the representations?

Designing Tasks and Questions for the Interview Protocol

After selecting the concept, the interviewer determines tasks and questions that will be used for the interview protocol. During a class session, the instructor shows teacher candidates samples of interview protocols and follow-up questions. For example, an interview protocol with fraction tasks for a young child may include tasks where the child shows one-half or one-fourth by using various concrete manipulative region models, a set model with different numbers in the set, or a string or long strip of paper as a length model. Another task may require the child to color a drawing or make a drawing of a fraction amount or write symbols that show fraction amounts. These tasks would be followed up with questions asking the child to explain the process used to arrive at the response and to generalize the concept by comparing the different fraction models. For example, one-half of a circle and one-half of a set of 16 snap cubes have a very different appearance. An important question to ask the child would be, " Why are both quantities called one-half?"

Class discussion focuses on questions the interviewer prepares prior to the interview and those that might be used as follow-up questions to encourage children to explain their thinking. A number of researchers have focused on the importance of developing questioning skills in the teaching and learning of mathematics (e.g., Carpenter, Fennema, Franke, Levi & Empson, 1999; Mewborn & Huberty, 1999). In a study of 48 interviews conducted by teacher candidates, analysis of the questions used by the candidates included the following categories: (1) *Checklisting* — the interviewer proceeded from one question to the next with little regard for the child's response, (2) *Instructing rather than assessing* — the interviewer asked leading questions or taught the concept to the child, and (3) *Probing and follow-up* — the interviewer used questions to invite a further response from the child (Moyer & Milewicz, 2002). Examples of these question types are presented in Figure 2. An important point of these examples is to prompt teacher candidates to ask questions that encourage children to make connections and synthesize new ideas.

Example #1 – Checklisting

Teacher:	What part is shaded?
Child:	The left.
Teacher:	How much is shaded?
Child:	The whole.
Teacher:	How much is shaded in this circle?
Child:	Half. (Grade 2 interview)

Example #2 – Instructing Rather than Assessing

Teacher:	So what did you have to think about to know that? You had to think about it as having what? You had to think about the whole thing, the whole set, you had to think about it kind of being in four different groups and then you could take one of them, right?
Child:	Yes. (Grade 1 interview)

Example #3 – Probing and Follow-Up Questioning

Teacher:	You said that this is half of a circle, this is half of a square, and this is half of the counters. How can they all be called half when they don't look the same?
Child:	Because this is red, this is red and this is white.
Teacher:	But look at the shapes. This is just a piece of the circle and a piece of the square, but this is 6 counters. How can they all be one-half?
Child:	Because you have two and they're cut.
Teacher:	We cut the paper circle and square, but we didn't cut the counters.
Child:	They're not the same, but they are the same....because this is two and this is two and this is two.
Teacher:	What do you mean 'this is two'?
Child:	They all have two parts. They all have partners. (Grade 2 interview)

Figure 2. Question types used by teacher candidates

The design of these tasks and questions provides the instructor with important information about the teacher candidate. For example, what tasks and questions are selected by the prospective teacher and do they match the concept being assessed? Does the prospective teacher develop questions and tasks that differentiate and provide extensions depending on the child's responses to the interview questions? The teacher candidate's selection of tasks and questions in developing a profile of the child's understanding of the concept are indicators of the teacher candidate's framework for the mathematics concept itself.

After designing the interview plan, teacher candidates submit this information to the instructor and receive feedback on the alignment of the concept, grade level, and tasks and questions selected for the interview. Candidates receive specific suggestions and feedback to strengthen the interview plan. These suggestions often focus teacher candidates on ordering and preparing a script for the tasks and questions to be asked, identifying alternative tasks and questions based on possible responses, and preparing questions to follow-up typical responses. This "what if" thinking helps teacher candidates consider what to say and do in a variety of situations prior to conducting the interview.

Conducting the Interview and Gathering Student Data

Conducting the Interview Protocol

Teacher candidates gather all of the materials necessary to conduct the interview, including a tape recorder, and select a quiet place for the interview so that the child will not be distracted. The interviews are planned for approximately 20-30 minutes, although some interviews are shorter or longer depending on the interviewer and the child. Interviewers use strategies to set the child at ease during the interview, such as beginning with questions at a low level of difficulty and avoiding evaluative statements that may discourage the child. When a child gives an incorrect response, the interviewer moves on, rather than stopping to teach the concept. The interview is a time to gather information, not to teach the concept to the child. For the teacher candidate, an important goal of this process is to focus on learning how young children think and to practice assessment, observation, and questioning techniques.

Gathering Data

During the interview it is important to document as much of the interaction as possible. Interviewers audiotape and prepare an abbreviated transcript of the interview, which they use to analyze their own questioning skills. They take notes on the child's behaviors, gather the child's work, and may photograph parts of the interview. Their purpose in gathering this data is to document the interview and provide evidence that supports their evaluation of the child's knowledge during analysis of the interview.

Asking Good Questions

The verbal interaction with a child, followed by a critical analysis of the interview transcript, allows teacher candidates to examine their own verbal interactions with children and identify examples and nonexamples of good questioning from their own transcripts, using the categories in Figure 2 as a guide. The instructor evaluates teacher candidates on the types of questions they prepare and also those that they create "on-the-spot" during the interview, looking for variety in the interviewer's questions. Candidates evaluate instances in which the child provides a response and the interviewer uses a specific follow-up question. These follow-up questions are evaluated for the interviewer's attention to the child's responses and probing for additional information or deeper understanding.

In addition to teacher candidates using the transcript for a final interview report, the questions in the transcript are used during an in-class discussion session. Members of the class examine the types of questions they used during the interviews and the frequency of those questions to identify their own questioning routines. This self-examination is effective in focusing teacher candidates on the types of questions they use (Moyer & Milewicz, 2002).

After the Interview

After teacher candidates conduct the interview with the child, they gather information and analyze their findings in a final report using the guidelines in Appendix A.

Analyze the Data and Evaluate the Child's Knowledge

To prepare the final report, teacher candidates synthesize all of the information from the interview. They analyze the gathered data, including the interview transcript, the child's writings and drawings, their notes, and any photographs. They review the data and present a holistic picture of the child's mathematical knowledge. During the analysis, teacher candidates form opinions about the child's knowledge and base these opinions on different sources of evidence from the interview. They use specific mathematics terminology and prepare a report similar to the formal documentation of a child's academic school record. Being able to describe a child's understanding of mathematics content, supporting that description with evidence, and articulating the information using professional language and terminology are important skills that classroom teachers use during the mathematics evaluation and assessment of children.

Develop an Individualized Instructional Plan

The information gathered from the assessment is then used to develop future instruction. In this part of the report, teacher candidates design a plan for instruction based on the child's understanding of the concept. They are encouraged to go beyond general comments in the instructional plan and be specific in their suggestions. For example, if teacher candidates assessed basic division concepts, they may suggest that the instructional plan for the child should include more manipulatives. This may be an important teaching strategy, but it is too general. Rather, the plan should be more specific about why and how manipulatives might be used to address the concept as illustrated in Appendix A. Developing an instructional plan helps teacher candidates focus on specific mathematical concepts and strategies that support student achievement, rather than on less effective generalizations.

Reflect on the Interview Process

The final requirement of the assignment is a reflection on the interview process. Teacher candidates report on the length of the interview, indicate what they learned about interviewing techniques, and discuss changes in future interviews that might improve the interview. They are also asked to think about what they have learned through the interview process about children's learning and mathematical thinking. Some teacher candidates comment that they are surprised by the depth of mathematical knowledge of the children they interview. In some cases, a child who was quiet or considered "average" during classroom interactions surprises the interviewer with a wealth of knowledge and understanding about a topic. Teacher candidates who have this experience often report that they will use various evaluation methods in their own teaching and not make assumptions about children's knowledge without using several assessment methods. The experience also illustrates that paper-and-pencil methods of assessment in mathematics do not always reveal a complete picture of children's mathematical knowledge.

The Evaluation Rubric

The evaluation rubric used with this assignment has been through several iterations. The current rubric is in Appendix B. The center portion of the rubric, with scores from 2-4, allows for a differentiation among those students who minimally meet the requirements of the assignment from those who meet the requirements thoroughly.

During the 2002-2003 academic year, the reliability of the rubric to assess the interview reports was determined. Five sample papers (two high papers, two middle papers, and one low paper) from a previous class were used to set a standard for the scoring rubric. The first reader was the course instructor. The second reader was a master elementary classroom teacher. Using a class set of assignments from 25 teacher candidates, all of the papers were independently scored by the two raters. The intraclass correlation coefficient used to compare interrater reliability was 0.97. The two scores for each report were averaged to determine one composite score for the report. Scores had a range of 0 to 5, with 5.0 indicating the highest score possible.

The 25 composite scores earned by the teacher candidates in this class had a range of 2.325 to 5.0, a mean of 3.8, and a standard deviation of 0.74. Elements of reliability and validity evident in the assessment include consistency with teacher candidates' scores on other tasks, consistency with coursework in other classes, alignment with standards, and differentiated range of scores.

Conclusion

Using alternative forms of assessment in mathematics, such as interviewing, has grown in popularity as a result of new mathematics standards (NCTM, 2000) and calls for assessment reform (Huinker, 1993; Stenmark, 1991). Teacher education courses have changed as a result of these shifts to require elementary teacher candidates to develop principled knowledge of mathematical concepts (Ball, 1991) and an understanding of how students think and reason mathematically. This assignment provides prospective teachers with a practical experience that allows them to develop and demonstrate their knowledge and skills in mathematics in their preparation for teaching. It also provides university instructors with another means to evaluate their future teachers.

References

Ball, D. (1991). Research on teaching mathematics: Making subject matter knowledge part of the equation. In J. Brophy (Ed.), *Advances in research on teaching*, Vol. 2 (pp. 1-41). Greenwich: JAI Press.

Buschman, L. (2001). Using student interviews to guide classroom instruction: An action research project. *Teaching Children Mathematics, 8*(4), 222-227.

Carpenter, T. P., Fennema, E., Franke, M. L., Levi, L., & Empson, S. B. (1999). *Children's mathematics: Cognitively guided instruction.* Portsmouth, N.H.: Heinemann.

Chappell, M. F., & Thompson, D. R. (1994). Modeling the NCTM *Standards*: Ideas for initial teacher preparation programs. In D. B. Aichele & A. F. Coxford (Eds.), *Professional development for teachers of mathematics* (pp. 186-199). Reston, VA: National Council of Teachers of Mathematics.

Cross, L., & Hynes, M. C. (1994). Assessing mathematics learning for students with learning differences. *Arithmetic Teacher, 41*, 371-377.

Fennema, E., & Carpenter, T. P. (1996). A longitudinal study of learning to use children's thinking in mathematics instruction. *Journal for Research in Mathematics Education, 27*(4), 403-434.

Fennema, E., Franke, M. L., Carpenter, T. P., & Carey, D. A. (1993). Using children's mathematical knowledge in instruction. *American Educational Research Journal, 30*, 555-585.

Huinker, D. M. (1993). Interviews: A window to students' conceptual knowledge of the operations. In N. L. Webb (Ed.), *Assessment in the mathematics classroom* (pp. 80-86). Reston, VA: National Council of Teachers of Mathematics.

Labinowicz, E. (1987). Assessing for learning: The interview method. *Arithmetic Teacher, 35*, 22-25.

Long, M. J., & Ben-Hur, M. (1991). Informing learning through the clinical interview. *Arithmetic Teacher, 38*, 44-46.

Mewborn, D. S., & Huberty, P. D. (1999). Questioning your way to the *Standards. Teaching Children Mathematics, 6*(4), 226-227, 243-246.

Moyer, P. S., & Milewicz, E. (2002). Learning to question: Categories of questioning used by preservice teachers during diagnostic mathematics interviews. *Journal of Mathematics Teacher Education, 5*(4), 293-315.

Moyer, P. S., & Moody, V. R. (1998). Shifting beliefs: Preservice teacher's reflections on assessing students' mathematical ideas. In S. B. Berenson & K. R. Dawkins (Eds.), *Proceedings of the Twentieth Annual Meeting of the North American Chapter of the International Group for the Psychology of Mathematics Education* (Vol. 2, pp. 613-619). Columbus, OH: ERIC Clearinghouse for Science, Mathematics, and Environmental Education.

National Council of Teachers of Mathematics. (2000). *Principles and standards for school mathematics*. Reston, VA: Author.

Peck, D. M., Jencks, S. M., & Connell, M. L. (1989). Improving instruction through brief interviews. *Arithmetic Teacher, 37*, 15-17.

Reys, R. E., Lindquist, M. M., Lambdin, D. V., Smith, N. L., & Suydam, M. N. (2004). *Helping children learn mathematics* (7th ed.). Hoboken, NJ: John Wiley & Sons, Inc.

Stenmark, J. K. (1991). *Mathematics assessment: Myths, models, good questions, and practical suggestions*. Reston, VA: National Council of Teachers of Mathematics.

Patricia S. Moyer-Packenham, Associate Professor of Mathematics Education in the Graduate School of Education and Director of the Mathematics Education Center (http://gse.gmu.edu/centers/cscvm/main.html) at George Mason University, teaches graduate mathematics courses in the Mathematics Education Leadership and Elementary Education programs. She received her Ph.D. in Curriculum and Instruction with a specialization in mathematics education from the University of North Carolina at Chapel Hill. Her research focuses on uses of mathematics representations and teacher development in mathematics. Her publications include the book, *What Principals Need to Know About Teaching Mathematics*, journal articles, and contributions to mathematics methods textbooks. [pmoyer@gmu.edu]

Appendix A.
Student Assessment Interview Report Guidelines

Student Work Samples

Collect and document three different forms of representation (concrete, pictorial, abstract symbols) during the interview to elicit the child's level of understanding. The report must include samples of the child's computations, writings or drawings, as well as a description of how the child used concrete objects during the interview or photographs of the child's work.

Question & Response Interview Transcript

Audiotape the interview. (Be sure to ask the child's teacher and parent for permission.) Type a basic transcript of the interview from the audio recording. Type only those questions and responses that pertain to mathematics. Be sure to include your questions and the child's responses. Indicate what you said and what the child said by using T for you (the teacher) and C for the child.

Questioning Competence

The questions and follow-up questions that you use during the interview will be evaluated. You will be evaluated on the *quality* and the *types* of follow-up questions you use during your interaction with the child. Your textbooks and readings provide direction on the types of questions that are appropriate in an interview and that go beyond factual information to deeper understanding.

Evaluation of the Child's Mathematical Knowledge

Write an evaluation of the child's mathematical knowledge in the content area. Use evidence from the interview to support your conclusions. Use your textbook to help you describe the specific types of behaviors and verbalizations you observed using specific mathematical terms. For example, if you conclude that the student has an understanding of addition of fractions with like denominators, you should base this on evidence that you present that shows the child was able to represent $\frac{3}{5}$ and $\frac{4}{5}$ with fraction pieces (concrete), and/or the child used a drawing to find the sum (pictorial), and/or the child computed the answer with symbols (abstract). Give specific examples of the child's responses to support your statements.

Instructional Plan

Develop a suggested instructional plan for the child. Your assessment of the child's thinking should give you some information for planning instruction. Your suggestions should be based on what you learned about the child during the interview. Many general suggestions can be valuable for children. However, your recommendations should relate to specifics. For example, if you assessed basic division concepts and you suggest that the instructional plan for the child should include more manipulatives, that would be an important teaching strategy, but it would be too general. You should be more specific about why and how manipulatives might be used. Example: "The student had difficulty making 3 equal groups from a set of 21 chips; therefore, the student should be given more experiences with grouping and partitioning manipulatives in sets of 15 to 30 to develop both the measurement and partitive concepts of division."

Reflection of the Interview Process

Comment on the interview process. How long did the interview last? What did you learn about interviewing techniques? What did you learn about your ability to create mathematics questions and tasks for this concept? If you were to conduct the interview with another child, would there be any changes in your questions, either the order or the level of difficulty, or the materials you had available for the child to use? Why or why not? What have you learned about how children learn mathematics from this interview? How might a classroom teacher use the diagnostic mathematics interview to assess children?

Appendix B. Rubric for Assessment Interview Report

Criteria	Exceeds Requirements	Meets Requirements	Needs Improvement	Inc.	Weight
	5	**4 3 2**	**1**	**0**	**x .05=**
Is the required information present about the <u>child</u> interviewed?	In addition to the required information, the Report includes information about the child's performance in other academic, social, or behavioral areas.	The Report includes the child's grade level, age, gender, race, academic ability level, and the child's level of understanding about the mathematics concept.	One or more of the required descriptive items about the child is missing.		
	5	**4 3 2**	**1**	**0**	**x .10=**
Has the interviewer selected one specific mathematics <u>concept</u> and assessed the concept using three different <u>forms of representation</u> (concrete, pictorial, abstract)?	Information on age-appropriate variations of the mathematics concept was gathered in preparation for the interview. One math concept is clearly described and mathematically accurate. Three different forms of representation, with different examples in each form, are designed for use in interesting and creative ways. Connections are made among representational forms.	One age-appropriate mathematics concept is selected, mathematically accurate, and clearly described. Three different forms of representation are described and used appropriately to assess the mathematics concept. Different examples may be used within each representational form.	One or more mathematics concepts are selected. They may not be age-appropriate. The Report is missing one or more forms of representation.		

Criteria	Exceeds Requirements	Meets Requirements	Needs Improvement	Inc.	Weight
	5	**4 3 2**	**1**	**0**	**x .15=**
Do the <u>tasks and questions</u> match the specific mathematics concept being assessed? Is there variety in the tasks and questions used for each of the three different forms of representation?	In addition to the tasks/questions being aligned with the math concept, there are questions that differentiate and provide extensions for different levels of student performance. In addition to the variety of tasks/ questions for each of the three forms of representation, tasks that show creativity and will be motivating for a child are included.	The tasks and questions designed for the interview are aligned with the mathematics concept being assessed. There are a variety of tasks and questions for each of the three forms of representation.	The tasks and questions designed for the interview are not clearly aligned with the mathematics concept being assessed. The Report is missing tasks/questions that address one or more of the forms of representation.		
	5	**4 3 2**	**1**	**0**	**x .10=**
Are the child's <u>work samples</u> included with three different forms of representation present in the work samples?	In addition to the variety of work samples from the child showing examples in each of the three forms of representation, a creative way of providing an explanatory overview of the child's work is included.	There are a variety of work samples from the child included showing examples in each of the three forms of representation (concrete, pictorial, abstract).	There is only one work sample in each of the three forms of representation or work samples from one form of representation are missing.		

Criteria	Exceeds Requirements	Meets Requirements	Needs Improvement	Inc.	Weight
	5	**4 3 2**	**1**	**0**	**x .15=**
Is the required question and response interview <u>transcript</u> present?	The Report includes a complete transcript of the mathematics interview that includes descriptive information on both the behaviors and the actual verbalizations that occurred during the interview.	The Report includes a transcript of the mathematics interview using the interviewer and the child's actual verbalizations from the interview (T for teacher, C for child).	The Report includes a transcript of the mathematics interview, but some parts of the interview conversation are missing.		
	5	**4 3 2**	**1**	**0**	**x .10=**
Do the initial and follow-up <u>questions</u> used by the interviewer demonstrate variety and higher levels of questioning? Are specific follow-up questions used appropriately?	The transcript shows that during the interview, the interviewer used a variety of questions to encourage the child to express his/her thinking, used many higher-level questions to encourage deeper thinking and responses from the child, and used specific follow-up questions to probe for understanding.	The transcript shows that during the interview, the interviewer used a variety of higher-level questions to encourage deeper thinking and appropriate follow-up questions to probe for understanding.	The transcript shows that during the interview, the interviewer used very few probing and follow-up questions when a specific follow-up question would have been appropriate.		

Criteria	Exceeds Requirements	Meets Requirements	Needs Improvement	Inc.	Weight
	5	**4 3 2**	**1**	**0**	**x .15=**
Does the evaluation accurately represent the child's current level of understanding on this concept using supporting evidence and work samples from the interview?	The evaluation provides an accurate and detailed description of the child's current level of understanding on the concept. Many different and specific examples from the interview are given, including the child's quotations, student work, and information from other sources on math development, to provide supporting evidence for the evaluation of the child.	The evaluation provides an accurate description of the child's current level of understanding on the mathematics concept. Different examples from the interview are given, including the child's quotations and student work, to provide supporting evidence for the evaluation.	The evaluation provides a minimal description of the child's understanding on the mathematics concept. A few examples from the interview are given, but there is not enough information to provide supporting evidence for the evaluation.		

Criteria	Exceeds Requirements	Meets Requirements	Needs Improvement	Inc.	Weight
	5	4 3 2	1	0	x .10=
Does the instructional plan prescribe developmentally appropriate next steps for instruction and take into account the child's current level of understanding on this concept?	The plan is a creative, detailed description of developmentally appropriate next steps for instruction taking into account the child's current level of understanding. The plan identifies many specific examples of activities and tasks that would further enhance this child's knowledge of this concept. Information from other sources on math development and child development was used.	The instructional plan describes developmentally appropriate next steps for instruction. The plan identifies several specific examples of tasks that would be appropriate to further enhance this child's knowledge on this concept. The plan describes these tasks in relation to the child's current level of understanding.	The plan describes some next steps for instruction that may not be developmentally appropriate. The plan gives general (rather than specific) examples of activities and tasks for the child. The tasks may not be appropriate either for the child or the development of the math.		
	5	4 3 2	1	0	x .10=
Is there an appropriate reflection and evaluation of the interview process?	In addition to the required information, the Report includes a detailed analysis, self-reflection, and self-evaluation of the interview process.	The Report includes a reflection and evaluation on the interview process including the required elements.	The Report does not include one or more of the required elements for the reflection.		
TOTAL SCORE					

Roth McDuffie, A., Mather, M., and Reynolds, K.
AMTE Monograph 1
The Work of Mathematics Teacher Educators
©2004, pp. 189-204

12

Professional Development for Teacher and Teacher Educator Through Sustained Collaboration

Amy Roth McDuffie
Martha Mather
Karen Reynolds
Washington State University Tri-Cities

During a year-long professional development project, a mathematics teacher educator collaborated with a middle-school mathematics teacher. The collaboration involved co-planning, implementing, and analyzing lessons with a focus on problem-based tasks and communication. Through this sustained and in-depth collaboration, the teacher experienced transformative learning and demonstrated clear shifts in her practice. The teacher educator also experienced growth by sharing in analyzing and developing solutions to dilemmas encountered in teaching and learning from a Standards-based perspective in a challenging context. The teacher educator's practice was enhanced through sharing these lessons with her university students.

Mathematics educators and researchers have called for increased efforts to situate professional development in the context of teachers' practices. In doing so, teachers have more opportunities to work through problems central to teaching and learning (Putnam & Borko, 2000; Smith, 2001; Thompson & Zeuli, 1999). Smith (2001) argues that professional development needs to be sustained and related to the teaching cycle (i.e., planning, teaching, and reflecting). Considering these perspectives, I[1] embarked on an in-depth, year-long professional development and research project in which I

---

[1] Although this manuscript has three authors, the pronoun "I" is used, referring to Roth McDuffie, because the manuscript focuses on the collaboration between Roth McDuffie and Ms. Lerenz. Mather and Reynolds contributed to this work during data analysis and writing.

collaborated with a middle-school mathematics teacher, Ms. Lerenz (a pseudonym), as her teaching became more *Standards*-based. My initial intent was focused on supporting Ms. Lerenz; however, I found that the collaboration also facilitated my growth as a teacher educator.

Although my purpose in this paper is to reflect on our professional development efforts, the research methodology merits explanation. Rather than considering research *on* the professional development process, I investigated Ms. Lerenz' practice using research as *part of* the professional development process by documenting the nature of our work and the teacher's practice (Simon, 2000). Thus, my methodological approaches complemented our collaboration rather than created barriers to interactions.

Participants and Context

The Teacher

Ms. Lerenz had been teaching middle-school mathematics for nine years. Her responses to a pedagogical beliefs interview conducted before our work began (using a protocol adapted from Peterson, Fennema, Carpenter, and Loef [1989]) and my interactions with her indicated that she believed students brought many ideas and strategies to mathematics learning, and that a teacher's role was to provide meaningful opportunities for students to build understandings. In sum, Ms. Lerenz held beliefs consistent with practices described in the National Council of Teachers of Mathematics (NCTM) *Standards* document (2000). However, Ms. Lerenz was not satisfied with her own efforts to enact these beliefs in practice. Throughout her career Ms. Lerenz regularly participated in local professional development opportunities (e.g., attending workshops and conferences) focused on developing her practice towards more effective implementation of teaching approaches that were consistent with her beliefs. In spite of these efforts, Ms. Lerenz was disappointed that her past experiences with professional development felt like one-shot workshops resulting only in inserting a new activity, without substantive change in her practice (Interview, August 01). This experience is consistent with Thompson and Zeuli's (1999) description of *additive* learning (adding a new skill to an existing repertoire) instead of *transformative* learning (making substantive changes in deeply held beliefs, knowledge, and habits of practice). Ms. Lerenz sought transformative learning.[2] Consequently, her primary motivation for engaging in our year-long collaboration was the hope of making meaningful and substantive changes in her practice, unlike her past efforts.

The Teacher Educator

I was in my fourth year as a mathematics educator at a university local to Ms. Lerenz' school. Prior to earning a Ph.D. in mathematics education, I taught middle- and high-school mathematics. Ms. Lerenz knew my reputation as a teacher educator in the local community and knew that I had been a mathematics teacher. That knowledge helped to establish trust and credibility with Ms. Lerenz even before we met. As we worked together and I shared personal teaching experiences and struggles, it became clear that Ms. Lerenz considered me a colleague, both because of my contributions in our collaboration and because of our common experiences with secondary teaching.

Just as Ms. Lerenz was dissatisfied with her lack of substantive changes in instructional practice, I entered our collaboration questioning the extent to which my past professional development efforts with teachers resulted in meaningful changes in their instructional practice. In supporting Ms. Lerenz, I sought ways to provide meaningful connections between research, theory, and practice, as recommended in the literature (e.g., Putnam & Borko, 2000; Thompson & Zeuli, 1999). My primary question was, "Would our sustained collaboration based in her classroom practice result in substantive and lasting change, going beyond the *one-shot* attempts that Ms. Lerenz and I had previously experienced?"

The School and Classroom Setting

Ms. Lerenz worked in a school that presented many complex teaching issues, especially with language and culture. This school had 1100 students in Grades 6 through 8 with 68% receiving free or reduced lunch, 62% racial/ethnic/language minority (primarily with Spanish as their first language), and 19% migrant families. My observations and discussions

---

[2] When I introduced Thompson and Zeuli's (1999) framework to Ms. Lerenz, she agreed that additive learning described her past experiences.

with Ms. Lerenz indicated that administrative and collegial support for improving teaching and learning was limited.

To maintain focus, we selected one of Ms. Lerenz' five mathematics classes that was classified as a "regular education," seventh-grade mathematics class. The school was using what most educators consider to be a traditional textbook (Silver, Burdett, & Ginn, 1994). Ms. Lerenz began the year with 26 students, and ended with 20 students, with changes throughout the year. Her class composition was in flux primarily due to students from migrant families leaving and returning to school and changes in classroom placements for English Language Learners (ELL).

Ms. Lerenz' Practice at the Beginning of the Year

Initially, most of Ms. Lerenz' lessons followed a traditional structure whereby she introduced a topic, provided examples, and had students practice problems similar to the examples. However, unlike a traditional approach, when introducing a topic, Ms. Lerenz attempted to involve her students through questioning, rather than simply telling them information. Approximately once a month, Ms. Lerenz invited me to observe a "problem-based learning" lesson. In the beginning of the year, Ms. Lerenz selected a task from an outside resource because she believed that her text did not provide true problem-solving tasks, and students worked on the task for two to three days. The task was inserted in the unit, in a manner similar to the implementation of new ideas after professional development experiences in the past.

For example, in October Ms. Lerenz selected the "Menu Task" (Hayes, 2000), her second problem-based task for the year. At this point in our collaboration, we had not yet begun to co-plan lessons, and thus, I was not working with Ms. Lerenz in analyzing and selecting tasks. For this task students received a restaurant menu with over 40 food and drink options and an order slip. The order slip contained seven lines for food items, and separate lines marked *subtotal*, *tip*, and *total*, respectively. The students responded to four prompts, each labeled for a day of the week and a scenario. On Monday the prompt read,

> Your family (which is your cooperative group) is going
> out to a restaurant for dinner. Use the menu to select

what you want to eat. Record your selections on the order form and add up the total cost. Remember to figure in a 15% tip for service.

1. The total cost for the meal was _____.
2. If the total cost for the dinner was evenly split between each member of the group, each person would pay _____. (Hayes, 2000, p. 6)

In administering the task, Ms. Lerenz circulated around the room, asking and responding to students' questions as the students worked in groups of three to four. The following dialogue reflects a typical exchange between Ms. Lerenz and a student. Although it is not evident in print, Ms. Lerenz spoke quickly, asking and then, without pause, answering her own questions. Notice that questioning became telling midway through the dialogue. Just prior to this exchange, Mario asked Ms. Lerenz for help starting on the "Monday" question on the Menu Task. Ms. Lerenz responded,

> What, what did it ask you to do? ... 'What are you going to eat for dinner, and how much is that going to be?' Right? And ... this is what you get if you go to the restaurant, right? [Pointing to the menu] ... The waitress writes it down right here [pointing to the order slip], and then she totals it up right there [pointing to the subtotal line]. That's what subtotal means. And then the tip is the 15% that you have to add on. And then that's the final total that you pay [pointing to "total"]. (October 9)

As Ms. Lerenz asked and answered these questions, Mario seemed tense and unsure. Once Ms. Lerenz began to tell Mario the steps to solve the problem (beginning with, "The waitress writes it down right here"), Mario listened closely, noting where to put his answers. In this episode, Ms. Lerenz attempted to facilitate learning through questioning as discussed in the summer workshop and earlier professional development experiences. However, she abandoned this approach in favor of telling, indicating that her practice did not match her beliefs.

Later in the class period, Ana asked Ms. Lerenz whether her work on the menu task "was right so far." Ms. Lerenz said, "How do you know if you're doing it right?... I can look at it afterwards if you want me to" (October 9). Again, Ms. Lerenz attempted to use a strategy discussed in the workshop, namely asking the student for an explanation of her understandings to encourage ownership. However, she shifted back to holding the authority for correct answers as she offered to "look at it afterwards." Directly and indirectly, Ms. Lerenz' interactions with students conveyed the message that she would explain the steps to solving a problem rather than having them develop a procedure and that she held the authority for correct answers.

Description of Collaborative Activities and Interactions

My work with Ms. Lerenz began when she enrolled in my two-week summer workshop for inservice teachers on problem-based learning and assessment. The workshop provided a common theoretical and research-based foundation for our subsequent work in her classroom in that we explored issues regarding selecting and using problem-based tasks. After this workshop, Ms. Lerenz agreed to continue working with me throughout the school year. In planning our collaboration, I endeavored to build our interactions around the teaching cycle of planning, teaching, reflecting (Smith, 2001). Not wanting to overwhelm Ms. Lerenz at the outset, I initially asked her only if I could observe her teaching and briefly discuss the lessons afterwards. After a few visits, I intended to ask her to meet to co-plan lessons and have more in-depth discussions after lessons. Unexpectedly, after the first two months, *Ms. Lerenz* requested an expansion of our work to include co-planning lessons and analyzing her practice in evening meetings. Her request suggested that she valued our work and was willing to commit her own time to further it.

Our work during the school year included (a) planning and reflec-tive discussions, focusing on problem-based tasks and assessment, questioning, and facilitating group work (7 meetings, 2-3 hours each); (b) classroom observations and video taping with discussions before and/or after class (24 observations, 1-2 hours each); (c) co-teaching lessons (3 lessons); and (d) regular email and phone conversations (over 30 correspondences).

I believed that Ms. Lerenz' learning would be most meaningful if it started with her perceived needs. Consequently, the most important aspect of my role in supporting Ms. Lerenz was listening and observing concerns that she identified as important. At times, listening involved finding patterns in her remarks and actions, and thereby helping her to identify her concerns. For example, just prior to expanding our collaboration in October, I had noticed that on three occasions Ms. Lerenz mentioned her struggles with forming good questions and facilitating student-to-student interactions during class discussions. I suggested to Ms. Lerenz that this seemed to be an emerging theme, and asked her if she would like to focus our work on this area both in addition to and as part of focusing on classroom-based assessment. Ms. Lerenz, in a somewhat surprised manner, agreed that she did keep returning to communication as an issue, and said she would like to pursue this focus further. In this way, I served as a mirror, reflecting back to Ms. Lerenz patterns and issues that she presented, but perhaps could not recognize without my presence.

In addition to serving as a mirror, at times I served as a lens from which to view her students' learning and her teaching. For example, during an observation in January, I noticed Marcos, a boy in the back of the room, was working well with his partner, explaining his ideas, and even scaffolding his partner's learning with his questions. However, Marcos did not contribute in class discussions, and became very quiet when Ms. Lerenz was nearby. I captured Marcos working on video, and showed it to Ms. Lerenz. After seeing and discussing the video, Ms. Lerenz began to bring out Marcos' strengths in small group work (e.g., Marcos served as the technology expert, helping students during a task involving a graphing calculator).

I also supported Ms. Lerenz by serving as a bridge to literature on teaching and learning. She often lacked time to select and read relevant research to inform her practice; however, when I offered readings that were pertinent to her concerns, she read and used the ideas in practice. For instance, during planning sessions focused on problem-based learning, I introduced the framework in Table 1 (next page) for evaluating performance tasks (adapted from Fuchs, Fuchs, Karns, Hamlett, & Katzaroff, 1999). Although we had discussed the features listed in the framework as part of the summer workshop, I had not

Table 1. Task Analysis Framework

Task Features	Tasks Analysis Score Results	
	Menu Task (Oct. 2001)	Cups Task (May 2002)
1. Requires students to explore important and grade-appropriate mathematics content.	1[1]	2
2. Contains a well-developed, meaningful problem context.	2	2
3. Requires students to solve at least two questions.	2	2
4. Provides students an opportunity to apply three or more grade-appropriate skills.	1[1]	2
5. Requires students to discriminate between relevant and irrelevant information.	1[2]	2
6. Requires students to generate information (e.g., more than one solution is possible).	1[3]	2
7. Requires students to explain their work.	0	2
8. Requires students to generate written communication (beyond explanations).	0	2
9. Incorporates multiple representations (e.g., tables, graphs).	0	2

Note: Table adapted from Fuchs et al., 1999. Scoring Levels: 2 – clear evidence of feature; 1 – some evidence of feature; 0 – no evidence of feature

[1]The lack of challenging, grade appropriate problems resulted in a score of 1 for Features 1 and 4.

[2]Although the students did not necessarily use all information given on the menu, discriminating between relevant and irrelevant information was not problematic, and therefore only partially demonstrated.

[3]More than one solution was possible in creating an order. However, because there were not inherent problems in selecting different items for an order, we gave a score of 1 to this feature.

shared this framework as a tool for analyzing tasks. Using this framework, Ms. Lerenz and I examined tasks for the nine features. Ms. Lerenz soon applied this framework in planning, finding it helpful as she evaluated and tailored tasks to fit her students' needs.

Additionally, responding to Ms. Lerenz' concern about students' communication in mathematics, I provided Ms. Lerenz with an article on communication (Brendefur & Frykolm, 2000). We used these ideas to analyze video, focusing on Ms. Lerenz' questioning, and classroom interactions. Other research that I introduced included the QUASAR project findings (Stein, Smith, Henningsen, & Silver, 2000) that examined how teachers' task implementation affected students' learning. Whenever I introduced literature, I summarized the relevant ideas to Ms. Lerenz to focus her reading. I coached her on reading research for practice: read the abstract and purpose for an overview; examine results and discussion sections, often skimming in places; if the study seems relevant, return to the theoretical framework and methodology to investigate whether perspectives or instruments might be useful in analyzing her practice.

We used these resources in analyzing Ms. Lerenz' practice in post-lesson discussions and when viewing video. Analyzing Ms. Lerenz' practice through the lens of the Brendefur and Frykolm (2000) article helped to take our discussions away from the personal (i.e., what Ms. Lerenz did) and into the realm of inquiry on practice, leading to conjectures about teaching and learning (e.g., ways to help students develop high quality explanations). In addition to providing research-based literature, I shared my teaching resources with Ms. Lerenz from which she selected the "Stacking Cups" task described next.

An Episode Illustrating the Nature of Our Collaboration

During planning meetings in May, Ms. Lerenz wanted a task that introduced linear relationships. Although the district's guidelines required that she introduce relations and functions, the text's chapter on this topic did not approach linear functions as a family of functions with specific properties. Instead, it focused more on input and output of values (Silver, Burdett, & Ginn, 1994). Among other resources, I showed her tasks in *Navigating Through Algebra in Grades 6 – 8*

(Friel, Rachlin, & Doyle, 2001). After reviewing several tasks, Ms. Lerenz found the "Stacking Cups" task. This task began with the following prompt: "You have been hired by a company that makes all kinds of cups – foam hot cups, ..., and more – of different sizes. For each of the kinds of cups it makes, the company needs to know the measurements of cartons that can hold 50 cups. Your task is to provide this information" (p. 81). The task prompts students to make a table and a coordinate graph of the number of cups and the height of a stack of cups that the teacher provides to measure. Then students describe the variables and their relationship, predict the height of a stack of 50 cups, explain their prediction, recommend carton dimensions, and compare results for different kinds of cups.

We examined the task together, considering it against the framework adapted from Fuchs et al. (1999). As indicated in Table 1, we found that the task strongly exhibited all nine features of the framework. In planning to develop a lesson around the task, we considered possible challenges the students might encounter. Ms. Lerenz expected that many of the students would simply multiply the cup height by the number of cups to find a function, rather than accounting for the overlap of cup heights as they are stacked.

In implementing the task, Ms. Lerenz and her students demonstrated growth in teaching and learning with problem-based tasks. Ms. Lerenz consistently showed an increased use of questioning as a strategy for facilitating learning and assessment of learning. This questioning was more conversational, with students' responses becoming richer in mathematical explanations (i.e., she decreased her use of rapid questions with little response from students as illustrated in the menu task). The following classroom excerpt from the cups task on May 30 is representative of these changes. Ms. Lerenz walked up to a group and asked how they were solving the problem.

Javier: What I did was, I measured this [pointing to cup's lip] and it was half an inch.

Lerenz: Half an inch for_____?

Javier: For just this part [pointing to cup's lip, specifically].

Lerenz: For the lip? Okay.
Javier: That's pretty much all that's gonna be stacked up, right? 'Cause I put 25, because I multiplied that, $\frac{1}{2}$ inch, by 50, for 50 cups. It was 25. And then three more inches for the whole cup, and its 28 inches.

The next excerpt from this same class exemplifies how Ms. Lerenz moved toward sense-making and testing conjectures in order for students to determine the quality of their solutions rather than checking with the teacher. Additionally, her questions reflect how she anticipated students' approaches and assessed for understanding in dealing with the overlap of cups in determining a linear function.

Lerenz: So did you add that [the three inch portion of the cup - the body of the cup]? Would it be added or multiplied?
Sophia: I added that on.
Lerenz: ...Test your theories out. Do it with five [cups], because you can always measure five because you have five [the students have 5 cups with which to work]. Make yourself do it this way first [using the rule the students developed] and then test it out and see if you're right. Does that make sense? [Ms. Lerenz left the students to test their conjecture.]

Professional Development Outcomes

At the beginning of the year, our primary challenge was to integrate problem-based learning more fully in Ms. Lerenz' instructional practice because it occurred on only a few days per month. By the end of the year, problem-based learning was an integral part and driving force for her instruction with problem-based tasks used regularly. These tasks came from outside sources, often borrowed from my materials. Ms. Lerenz used her text only as a resource for homework practice. The quality and implementation of the tasks also improved, as evidenced by comparing the menu task and the stacking cups task (see Table 1). Although students' learning is not the focus of this paper, the excerpts of class discussion provide a window into the change for students as well: her students made shifts from viewing mathematics as doing exercises with single, correct answers to becoming more persistent problem solvers, engaging in problems with multiple paths and solutions.

For Ms. Lerenz, the following themes emerged: (1) by focusing on a specific class and aspect of practice (i.e., problem-based learning), Ms. Lerenz found that initiating substantive changes was manageable; (2) through this focus, Ms. Lerenz' growth spread to other aspects of instruction (e.g., questioning strategies); (3) by situating our analysis and planning in classroom practice, Ms. Lerenz connected and implemented ideas from current research and theory into her practice. After the year concluded, she perceived that, for the first time, she had made transformative changes in her instructional practice. Moreover, Ms. Lerenz said that, "By working with you, Amy, I pushed myself to try more things that I wouldn't have tried otherwise."

As a mathematics educator, I experienced growth through a full year of regular participation in a classroom with all of the pressures teachers face (e.g., changing practice without *Standards*-based curricula and support, external emphasis on test scores, etc.). I was challenged to analyze and assist in developing solutions to teaching dilemmas as they occurred in practice as I served as both a mirror and a lens to Ms. Lerenz. Through the process of reviewing and evaluating current research and theory for its applicability to Ms. Lerenz' context and coaching Ms. Lerenz on reading research, I focused on reading from a practitioner's perspective. This perspective is important for me to keep in focus to understand the needs of preservice teachers. Additionally, after seeing the importance of structured analyses of tasks in planning instruction, I dedicated more attention to task selection in class discussions and incorporated the task features in Table 1 in a problem-based lesson planning assignment for my preservice teachers, an assignment they evaluated as being very worthwhile.

This experience also provided current, locally-based teaching and learning experiences to share with preservice and inservice teachers, many of whom argue that problem-based learning with its significant reading requirements is not practical for students who are ELL. The examples I shared from Ms. Lerenz' class served as an existence proof that *Standards*-based teaching can be successfully implemented with real contextual barriers present. In sum, I meaningfully supported a teacher in her growth process while enriching my practice and experiencing my own professional challenges and growth.

Final Reflections

In reflecting on our work for teacher educators who are considering embarking on a similar effort, I first considered aspects that contributed to developing and sustaining the collaboration – a collaborative effort that developed into a partnership. It seemed to be important to listen to Ms. Lerenz' needs rather than push my agenda, and to allow time for the partnership to develop. Setting aside my goals and focusing on listening was difficult for me at times because I had definite ideas for the direction of our work (e.g., more emphasis on assessing student work, an area that did not receive much attention during the year). However, it is clear that Ms. Lerenz showed the most growth in the areas that she identified as a focus (e.g., questioning strategies), and she frequently told me that she valued having a colleague with whom to listen and discuss *her* dilemmas. Similar to students, Ms. Lerenz learned most when she had a self-identified need. Additionally, staying focused on analyzing instructional practice through the use of frameworks from the literature seemed to mitigate Ms. Lerenz feeling defensive as if she were being evaluated. I also intentionally used the pronoun "we" in discussing instructional practice (e.g., "We should consider using this task...") to share ownership, emphasize our shared goals, and avoid sounding like I was evaluating her. Forming this partnership helped to sustain and motivate our work. We both had a sense that we needed to invest in reading as well as preparing and analyzing lessons for the partnership. In short, our work was bigger than either one of us as individuals.

Both Ms. Lerenz and I perceived that she experienced transformative growth, unlike her past professional development experiences. In reflecting on the factors that contributed to this growth, the most important aspects seem to be that it was sustained, entirely situated in her practice, and part of a committed partnership. Undoubtedly, the fact that Ms. Lerenz was motivated and sought opportunities to change was a key part of this success. This project did not address how to motivate teachers, and we cannot assume that motivation alone will lead to transformative growth.

In future collaborations, I intend to make the following changes. First, I will secure permission from Ms. Lerenz' students to share the videos with other teachers, including in my university courses. Although I use videos from published professional development materials, having locally situated videos might provide instances of practice that are perceived as being more credible to my preservice and inservice teachers. Additionally, I will begin planning the collaborative work with the teacher prior to the start of the school year to establish goals and possible foci. In doing so, the teacher has the opportunity to consider changes to beginning of the year activities in which classroom norms and expectations are established, and more time is available to establish a collegial relationship.

On the surface it appears that significant time and effort were expended for the benefit of only one teacher and her students, but greater benefit was indeed realized. Not only did the collaboration contribute to my professional growth as a teacher educator, a year after the work was concluded, it is clear that the experience ignited further growth. Ms. Lerenz has continued in her professional development, further refining changes in instructional practices. Moreover, she has begun to take on a role of a teacher leader, supporting other teachers in their growth process. It seems that this work has the potential to be both transformative and generative in nature, extending well beyond the collaboration.

References

Brendefur, J., & Frykholm, J. (2000). Promoting mathematical communication in the classroom: Two preservice teachers' conceptions and practices. *Journal of Mathematics Teacher Education, 3*(2), 125-153.

Friel, S., Rachlin, S., & Doyle, D. (2001). *Navigating through algebra in grades 6 – 8*. Reston, VA: National Council of Teachers of Mathematics.

Fuchs, L., Fuchs, D., Karns, K., Hamlett, C., & Katzaroff, M. (1999). Mathematics performance assessment in the classroom: Effects on teacher planning and student problem solving. *American Educational Research Journal, 36*(3), 609-646.

Hayes, L. (2000, October/November). Thanksgiving week with the relatives. *Oregon Mathematics Teacher*, 17-19.

National Council of Teachers of Mathematics. (2000). *Principles and standards for school mathematics*. Reston, VA: Author.

Peterson, P., Fennema, E., Carpenter, T., & Loef, M. (1989). Teachers' pedagogical content beliefs in mathematics. *Cognition and Instruction, 6*(1), 1-40.

Putnam, R., & Borko, H. (2000). What do new views of knowledge and thinking have to say about research on teacher learning? *Educational Researcher, 29*, 4-15.

Silver, Burdett, & Ginn. (1994). *Mathematics: Exploring your world*. Morristown, NJ: Author.

Simon, M. (2000). Research on the development of mathematics teachers: The teacher development experiment. In A. Kelly & R. Lesh (Eds.), *Handbook of research design in mathematics and science education.* (pp. 335-360). Mahwah, NJ: Erlbaum.

Smith, M. (2001). *Practice-based professional development for teachers of mathematics*. Reston, VA: National Council of Teachers of Mathematics.

Stein, M., Smith, M., Henningsen, M., & Silver, E. (2000). *Implementing standards-based mathematics instruction: A casebook for professional development*. New York: Teachers College Press.

Thompson, C., & Zeuli, J. (1999). The frame and the tapestry: Standards-based reform and professional development. In L. Darling-Hammond & G. Sykes (Eds.), *Teaching as the learning profession* (pp. 341-375). San Francisco: Jossey-Bass.

Amy Roth McDuffie, Assistant Professor of Mathematics Education at Washington State University Tri-Cities, is interested in preservice and inservice teacher professional development towards *Standards*-based practices. [mcduffie@tricity.wsu.edu]

Martha Mather, a graduate student in the Masters in Teaching program of the Department of Teaching and Learning at Washington State University Tri-Cities, is interested in classroom-based research of middle-level mathematics teaching and learning.

Karen Reynolds, an undergraduate student in the Department of Teaching and Learning at Washington State University Tri-Cities, focuses on reflection on practice to improve mathematics teaching and learning.

Schorr, R. Y.
AMTE Monograph 1
The Work of Mathematics Teacher Educators
©2004, pp. 205-222

13

Helping Teachers Develop New Conceptualizations About the Teaching and Learning of Mathematics[1]

Roberta Y. Schorr
Rutgers, the State University of New Jersey
Campus at Newark

This paper describes a professional development model designed to help teachers create instructional atmospheres in which all students have an opportunity to learn important mathematical concepts and processes with understanding. The model involves a sequence of activities in which teachers interact with students, other teachers, and teacher educators over time in the context of specially designed experiences. These interactions focus on thought-revealing activities. As teachers changed their practices, the nature of their students' thinking changed as well.

A central theme that permeates the recommendations of the National Council of Teachers of Mathematics (NCTM) is that "all students should learn important mathematical concepts and processes with understanding" (NCTM, 2000, p. ix). Teaching for understanding, however noble a goal, is easier said than done. In a study of fourth-grade teachers in one state, it was found that many teachers stated that they *wanted* to teach for understanding, and indeed, felt that they *were* teaching for understanding. Yet, despite their intentions, trained observers who visited their classrooms noted little evidence of practices that would encourage understanding. For example, teachers rarely probed students to determine whether their answers made sense; rarely asked students to explain, justify or share their reasoning; and usually focused their teaching on procedural knowledge, without encouraging deeper understanding (Firestone, Schorr, Monfils, 2004; Schorr, Firestone, Monfils, 2003).

[1] This work was supported, in part, by a grant from the National Science Foundation (#ESI-0138806). Any opinions, findings, and conclusions or recommendations expressed in this material are those of the author and do not necessarily reflect the views of Rutgers University or the National Science Foundation.

Several authors (Simon and Tzur, 1999; Simon, Tzur, Heinz, Kinzel, & Smith, 2000; Spillane & Zeulli, 1999) suggest that many teachers have a limited concept of the changes advocated by the *Standards*. Many interpret needed changes as discouraging telling and showing, using manipulatives, or having students work in small groups. Although these strategies may contribute to an enjoyable teaching environment, taken alone or in combination such strategies may not necessarily lead to the development of increased conceptual understanding in students (Schorr, Firestone and Monfils, 2003; Schorr and Koellner-Clark, 2003; Simon and Tzur, 1999; Spillane and Zeulli, 1999). These strategies may represent a step in the right direction, but the authors of the *Principles and Standards for School Mathematics* caution that "some of the pedagogical ideas from the NCTM Standards...have been enacted without sufficient attention to students' understanding of mathematics content" (NCTM, 2000, pp. 5-6).

According to Goldsmith and Schifter (1993), developing a better form of teaching requires more than the acquisition of some new instructional techniques or strategies. It requires a re-conceptualization of the entire teaching and learning process. Indeed, approaches that emphasize student understanding place demands on teachers to have a deep understanding of the content and how students learn the content (Ball, 2001; Cobb, Wood, Yackel & McNeal, 1993; Davis, 1992; Davis & Maher, 1997; Fennema, Sowder, & Carpenter, 1999; Franke and Kazemi, 2001; Lesh & Doerr, 2003; NCTM, 2000; Schorr, 2000; Schorr & Lesh, 2003; Simon & Tzur, 1999; Sowder & Philipp, 1999). In the absence of such understanding, it is difficult for teachers to make sense of students' explanations, probe for justifications, base instruction upon students' thinking, and promote an instructional atmosphere and culture that supports children as they explore, explain, defend, and reflect upon rich mathematical ideas (Davis, 1992; Davis & Maher, 1997; Schorr, 2000; Schorr & Lesh, 2003; Shafer & Romberg, 1999).

Teaching practices are not easily modified or changed. Teachers have well-developed ideas about the teaching and learning process as a result of their own experiences as a student and a teacher, such as their knowledge of the mathematical concepts involved, how children learn the concepts, how and when to assess students, what students can and cannot learn, and appropriate pedagogical approaches that should be used to teach concepts (Schorr & Koellner-Clark, 2003). When teachers do adopt specific changes or strategies into their classroom practice, they often do so within the framework of their older teaching ideas, many of which are "teacher-centered" (Cuban, 1993). Thus, they use manipulatives, but in very prescribed ways; students work in small groups, but solve routine problems; and, teachers ask students to explain their thinking, but settle for superficial responses and do not delve deeply into why and how the student came to a solution. Indeed, given the long history of relatively stable instructional practices in this country (Cuban, 1993), and teachers' often limited understanding of the mathematical content and associated pedagogical practices (Ma, 1999), it is not surprising that the changes reported in mathematics instruction are more akin to the adoption of new strategies or techniques rather than fundamental changes in practice (Schorr & Firestone, 2004).

General Components and Assumptions of the Teacher Development Design

In an effort to help teachers build a deep and substantive knowledge base about *Standards*-based teaching, this paper reports on a professional development design that provides a context for helping teachers revise, refine, and extend their ways of thinking about teaching and learning mathematics. The overall teacher development model involves a sequence of activities in which teachers interact with students, other teachers, and teacher educators over time in the context of specially designed experiences. These interactions are focused on thought-revealing activities.

Thought-revealing activities have several characteristics that make them different from problems that frequently appear in texts, on tests, or in workshops. For instance, they are designed to encourage participants to make sense of a situation based upon extensions of their personal knowledge and experiences, rather than conform to someone else's (i.e., the author's, the teacher's, or the teacher educator's) conception of what the right answer *should* be. They are also designed so that participants can judge for themselves whether or not the answer or product that they are supplying is good, and whether or not it is generalizable beyond the specific situation for which it was developed or used. In addition, thought-revealing activities are intended to produce trails of documentation that highlight important aspects about the constructs that develop for both students and their teachers. So, teacher educators can continuously adapt, modify, test, and extend their *own* thinking about the effects of an intervention, and consequently make informed decisions about the types of thought-revealing activities that would be useful for teachers and students in the future.

Thought-revealing activities are particularly useful because they accomplish several things simultaneously. First, they positively influence the nature and type of tasks that are offered to students as well as the manner in which these tasks are used within classrooms. Second, they provide an opportunity for teachers to reflect upon the effects of their efforts. Third, these activities allow teachers to use their own teaching experiences as the basis for building their own content knowledge and professional growth. The key component, however, is that all activities help teachers to reveal their thinking as well as test, extend, and share their thinking.

This professional development design has several levels. One level focuses on teachers' interpretations of students' developing interpretations of mathematical ideas. (See Lesh, Hoover, & Kelly 1992 for a more complete description of thought-revealing activities for students.) Students' solutions highlight the development of mathematically significant ideas involving concepts such as fractions, ratios, rates, proportions, or other important mathematical constructs. A second level focuses on teachers' developing conceptions about the mathematics itself, the ways in which students build understanding of the mathematics, and the teaching and learning process. A third level focuses on teacher educators' developing conceptions about the nature of teachers' developing knowledge and abilities. At each level, the students, teachers, and teacher educators are challenged to extend their ways of thinking and to develop in directions that they could judge to be continually better — even in the absence of a predefined conception of "best."

More specifically, student level thought-revealing activities create the need for students to construct, refine, and extend significant mathematical ideas (e.g., proportional reasoning, statistical reasoning, algebraic reasoning). (See Lesh, Amit, and Schorr (1997) for an example of statistical reasoning.) The problems encourage students to make sense of a situation by extending their personal knowledge and experiences. Students are required to reveal how they are thinking about the situation (including the givens, goals, possible solution paths, etc.) and the types of mathematical objects, relations, operations, patterns, and regularities that they are considering. In this way, students, teachers, and teacher educators are able to understand the nature of the mathematical thinking that is being considered.

Teacher-level thought-revealing activities revolve around students' responses to problem-solving activities to help teachers increase their knowledge base about *Standards*-based teaching, and alter their views of teaching and learning. Three examples of teacher-level thought-revealing activities and the corresponding teacher educator responses follow.

- Teachers generate observation guidelines that they and their colleagues use to observe students solving the problem activities. These observation guidelines reveal those things teachers want to observe as their students are solving problems (e.g., different strategies that may emerge, common misconceptions, different representations that may be useful). Having teachers share these guidelines with their peers and teacher educators and test them in their classrooms enables teachers to determine if all the important behaviors are in the guidelines. As teacher educators gain insight into the ways in which teachers view their students, they can use additional teacher-level thought-revealing activities to help teachers become better at observing their students' ways of thinking.

- Teachers share student products that display a particular type of mathematical reasoning or thinking and use the products to document different ways of thinking that can emerge as students solve structurally similar problems. In this way, teachers share what they consider to be the important mathematical ideas revealed in their students' work. Teacher educators can then use these products to prompt conversations about the mathematical ideas that emerged, or might have emerged, and how the ideas and representations are linked. In a closely related activity, teachers share videotapes that document an informative aspect of a teaching or learning episode. Teacher educators can use these episodes to consider the ways in which teachers respond to particular situations, and the practical implications of those decisions.

- Teachers consider how their students' thinking changes over the course of solving a problem or series of problems. This activity helps teachers focus on the evolution, development, and change in the way that a student thinks about a problem over time. Teacher educators and teachers can then discuss different learning trajectories and cycles that were observed, and hypothesize about student solutions involving structurally similar problems. Teachers can then test these hypotheses in real classrooms.

There is a critical difference between the approach described here and other approaches that use student thinking as a basis for teacher development. These particular activities go beyond having teachers think about student thinking in the abstract or in the particular case under scrutiny by helping teachers and teacher educators design tools for practice and explicitly think about their *own* thinking about student thinking. Further, teacher educators can use the results of these activities to make informed decisions about how to proceed in a professional development setting. At times, ideas or activities that seem to hold promise may not produce the desired results, at least not immediately. For example, when teachers first begin to generate observation guidelines, they often focus on general aspects of their students' work and interactions, such as whether or not all students are on task, or the roles that individual students are assigned within the group. The teacher educator would then ask teachers to use these guidelines during their own problem solving sessions with their peers and during actual classroom sessions. As teachers use these observation guidelines with their students, they are "testing" whether or not the guidelines are a useful tool to gain insights into their students' thinking. As these guidelines are shared with others, additions, deletions, or modifications can be made and the guidelines can be retested in classroom situations. The process continues and a trail of documentation about the types of things that teachers believe to be important is made. The observation guidelines become "tools" for both teachers and teacher educators to consider change. If the observations are not becoming progressively better, then the teacher educator must reconsider the nature of the professional development activities and the type of thought-revealing activities that are used. For example, teacher educators can combine information from the observation guidelines with examples of student products to trigger teacher conversations about the mathematical thinking that emerged. As teachers continually focus on what their students are doing, they become better at analyzing their thinking, and the cycle continues.

Specific Components of the Teacher Education Design

As part of this professional development design, teachers are asked to solve a series of thought-revealing activities that could later be used with their own students. They are then asked to share ideas with each other and with teacher educators about the main mathematical ideas, notations that are used or invented, representational systems that are employed or could be employed, and how these aspects are related. Teachers also consider proposed implementation strategies, pedagogical approaches, connections with other mathematical topics, and anticipated student outcomes. After implementing these activities in their own classrooms, teachers share the results of the implementation, including their observation guidelines and their students' work on the activities, their own thoughts about their students' mathematical ideas, the representations and notations that the students invented or chose, and their assessments of their students' work including relevant artifacts. Teachers also share implementation issues.

A key aspect of this design is that teachers are always asked to test all ideas in the context of their own classrooms. Thus, the discussions that take place at workshops are based upon their personal experiences with their own students. This is dramatically different from activities in teacher development workshops in which teachers solve a series of problems, often in a reform style, and are then expected to use the activities in their classrooms in a reform style. Missing is the deep reflection that is needed so that teachers can consider the impact on student thinking and understanding.

In this design, the development of teachers and their students tends to be highly interdependent. For example, when teachers first become involved in the process, the responses that students produce to the problem activities may not be as mathematically sophisticated as those produced later. As teachers develop new ways of implementing these activities and building upon the mathematical thinking of their students, the responses that students produce become qualitatively better. As the responses become better, the teachers continue to refine and revise their ways of thinking about their students' work, as well as the teaching and learning process.

A fundamental premise of this design is that the development of students and teachers is not simply the result of passive acceptance of information. Telling teachers about students' thinking is no more effective than telling students about a complex mathematical idea. The potential for development occurs when learners have the opportunity to build, revise, share, test, and extend their ideas as a result of their own experiences. In addition, teacher educators tend to be integral parts of the systems they are trying to understand and explain; they are also revising, extending, and refining their ways of thinking about how to use all of the information to design new and more appropriate learning experiences. Thus, teacher educators are not external to the process but are inextricably intertwined in it because they must also test their ideas with actual teachers, classrooms, and students. Teacher educators are attempting to determine if the teachers become better at interpreting students' thinking and using their understandings to inform instruction rather than merely using strategies or instructional sequences in reform styles.

Some Illustrative Examples

Consider an activity in which teachers share what they consider to be important things to observe as their students are engaged in solving problems in small group settings. In the beginning of one project, a group of urban sixth-grade teachers developed a list which included the degree of engagement on the part of the students (i.e., were all students participating?), the social interaction of students (i.e., were they all getting along?), the type of participation (i.e., who was doing what), and the types of skills that the students used (i.e., what procedures were they using?). Absent from the list was attention to the mathematics itself beyond the use of procedures. The teachers did not include any of the ways in which different procedures, concepts, or models were used, and how they related to each other. They did not include attention to different representational systems that the students may have developed or chosen, and how they related to each other. They also did not include attention to the ways in which ideas were generated and used over time. As the project progressed, the teachers became less focused on the social dynamics of how the students were working, although these still remained important; rather, they focused more on the mathematical thinking that emerged, and how that thinking changed over the course of the problem-solving experience.

As the observations changed, the teachers noted that their students were doing things that surprised them. For example, consider the following thought-revealing problem activity involving a comparison of offers from three different CD companies.[2] In Club 1, members initially got five CDs for $1 and were required to purchase five more CDs at the regular club price of $12.99 each over the next year. Club 2 offered 10 CDs for the price of one ($18.99) and members had to purchase six more over the next two years at the regular price. The third company offered to sell one CD at the regular price ($12.98) and a second at half price. Information was also provided regarding

---

[2] The problem activity, "CD Deals" along with the Zoo problem activity that appears in the next section were preliminary versions of activities involved in the PACKETS Investigations for Upper Elementary Grades, developed at Educational Testing Service with funding from the National Science Foundation.

shipping costs and other membership issues, such as cancellation policies. Students were asked to choose the company that offered the best bargain, to share their method of comparing clubs, and to develop a way for others to review their findings and consider new clubs.

As the teachers solved the problem, they noticed several methods for comparing the clubs and generalizing their processes for determining optimization, which in this case involved unit ratios. One teacher noted that her students decided that the best way to compare three CD club offers was to use the unit ratio, while minimizing or maximizing other costs such as shipping and handling. Indeed, this teacher noted that students were able to use the unit ratio with facility on a range of problems; in past years students had not been able to use unit ratios, even after the teacher had formally taught the concept.

Teachers also found that different students began to display talents and abilities. One particular fifth-grade teacher in a suburban community noted, "I had two classified students who had little confidence or success in traditional activities. Here they were able and willing to take chances. They shared their approaches and conversed with other students about the problems. Some of the so-called higher level students did not do so well, especially those students who, while great at computation, are impatient if the answer does not come quick and easy." One of her colleagues added, "there was one student who had been a constant problem in class. During these sessions he was totally involved. Although he is a special education student, he had real success here because he was able to explore and do the problems his own way."

These remarks were supported by dramatic changes in the quality of the students' work. Initial products tended to be aesthetically pleasing, but mathematically superficial. As the teachers became more interested in their students' thinking, their students tended to do more thinking. At the beginning of one project, a group of elementary teachers, who taught fourth, fifth, and sixth grades, solved a problem involving proportional reasoning; the goal was to create a school banner by scaling up a picture of a lion's face composed of various geometric figures (see Schorr & Lesh, 2003). A more elaborate story surrounded the problem, which included the need for detailed descriptions of the scaling process as part of the solution. The teachers solved the problem in a workshop setting with teacher educators present. As they shared solutions, the teachers noted that they had "used proportions to help scale-up the newspaper drawing to make the full banner." There was also discussion about how to implement the problem in the classroom. The teachers decided to allow small groups of children to work together to generate solutions to the problem without their intervention. In addition, teachers agreed to share student products and identify exemplary or illuminating examples of student work.

The teachers then implemented the activity in their own classrooms. Shortly after, they again met with teacher educators in a workshop setting to share classroom results. The following represents one example of a student product that the teachers identified as exemplary; all student products included detailed pictures of a lion made with several different pieces of colored construction paper:

> *Product 1: We made a little pattern for the lioness out of construction paper. Then we traced a cymbal for the head. Then we traced a plastic lid for the ears. For the eyes we traced the bottom of a glue bottle. For the nose we used a ruler. For the mouth we used a protractor.*

One teacher, reflecting the consensus of the group, noted that this was a "very good student product[s] because the students used many different objects to construct the lion head." She continued, "I really liked the way that the students were engaged in the activity, they used lots of different objects to construct the circles and other shapes." Throughout the entire discussion of the students' work, none of the teachers noted that the lions drawn by the students were not proportional to the original lion — even though the teachers had identified that characteristic as one of the most important mathematical ideas in the problem activity. The teacher educators asked the teachers to reflect upon the mathematics that they themselves had used and identified as being important (i.e., proportional reasoning) and to compare it with the mathematics that was evident in the student products. This discussion helped both the teacher educators and teachers to adapt their views regarding what was significant to the teachers in the teaching and learning process. As a consequence, teacher educators were better able to understand issues that were important to teachers and their students at a given point in time and deal with them in subsequent workshops and classroom visits. The teachers were then increasingly able to focus on the mathematical ideas of their students, and in turn, their students began to focus more on the mathematics as well.

Later in the project, a different problem activity was considered, solved, and discussed by the teachers in a workshop setting, again with the teacher educators present. The activity involved a map of an actual zoo, including a scale and the locations of various animals. The problem indicated that there were three classes who wanted to visit the zoo, each focusing on a different group of animals, with only two hours to complete the visit. They were given the three lists of animals chosen by each of the classes, including the amount of time each class wanted to spend at each animal exhibit. The assignment was to design a plan for each class, including a route the class would have to take and the amount of time needed to make the trip. In order to accomplish the task, students needed to explain the best way to plan the route for each class. If the trip exceeded the two-hour time period, they had to explain the best way to change the plans so that the class could see as many animals on their list as possible. Finally, students had to explain how they decided if a class had enough time to see everything on the list and how they chose the paths, so that the three classes could plan themselves the next time they wanted to visit the zoo.

The teachers solved the problem using several different methods. As with the lion problem, they shared their solutions and discussed the main mathematical ideas that they felt might be elicited, including units of time, operations with whole numbers, measurement using non-standard units, and optimization. Teachers then designed an observation list for observing their students, and outlined plans for implementation.

What follows is an example of work from one group of students:

Mrs. H, your class will make the 2 hours without any problems. Mr. H, your class will make the 2 hours with you having to sacrifice the [Amazon] exhibit. Mrs. B, your class will make the 2 hour time if you sacrificed seeing the antelope and the lions.

The way we figured out the problem is simple. First we had to pick the route, the route which we thought would be the shortest, and we would be able to see all the animals on your list. We picked the route by estimating which route would take less time and by rearranging the exhibits in a faster order. Next we had to measure the route. We measured the route with a piece of paper, which we measured on the scale. We also traced the route with measured lines and then counted the spaces between the lines. We counted up the spaces and that's how many minutes you would have to walk. Next, we added all the time you wanted to spend at the exhibits. We took both of the times and added them together. That was how many minutes your trip will take. If the sum was 120 minutes or less then you could see all the exhibits you wanted to see. If not, then we had 1 or 2 options. If the sum was over the limit of 120 minutes by 5 minutes or less then we just subtracted some of the time you wanted to spend at the exhibits. If the sum was over 5 minutes then we just eliminated a whole exhibit. You should see all of the animal exhibits at one side of the zoo then go to the other side instead of going back and forth. Sometimes you have to eliminate exhibits to save time. The way you do that is by picking the animals that are farthest apart and you wanted to study the longest.

Here are some tips you can use when planning a trip. You should predict how long it would take you to walk the route then pick what you're going to see. Try and pick a route that doesn't backtrack. If you eliminate an exhibit then try to pick the one that is farthest away from the rest of the exhibits planned and takes up the most amount of time.

Next time you plan a trip make sure that you use these steps to help you plan the best trip you possibly can. P.S. Plan ahead for bathroom breaks and resting periods.

At a follow up workshop with teacher educators present, the teachers shared ways in which different students in their classes had used measuring tools, such as string and paper, to determine the walking time. They also noted that some students had successively subtracted walking and/or visiting times from the total; others had dealt with walking and visiting times individually or by summing the total time for each as it occurred. Yet others had used a strategy involving multiplication; students assigned a number of steps to a given distance and then multiplied the number of steps by the time to walk a step to determine the elapsed time. The teachers were particularly impressed with students who "optimized" the time by rearranging the order of the visits to the animals, and then systematically developed a method to ensure that the trip could be made in the allotted time period. They also noted that the students had set parameters for ascertaining whether a given animal visit should be deleted or whether a visit should be decreased in time when the total time exceeded what was available. This group of teachers had clearly begun to notice mathematical thinking in very different ways than they had when their students solved the lion problem!

One teacher recommended that the students needed to become more reflective about their own work. She designed an "Assessing Myself" sheet for her students to use for self-assessment. She included several questions, noting that these questions were important for her students to consider so that they could become more reflective about their own problem-solving strategies, mathematical skills, and ideas. Two sample questions were:

- "What were you able to contribute to the solution of this problem?" and

- "If you could change your product, what would you do to make it better? Explain how these revisions might improve your products - feel free to use ideas from other groups, or ideas that you heard in your group that didn't get used."

The teachers continued to evolve. As they changed, the nature of teacher level thought-revealing activities changed as well. Although the teachers recognized that student-level thought-revealing activities were powerful, they realized that the activities represented only part of what should constitute a mathematics curriculum. Together, the teacher educators and teachers analyzed strands of mathematical content, and how they could be taught. The teachers began to implement these ideas in their own classrooms, again carefully observing their students and reflecting on their work.

Conclusions

When considering "understanding" in mathematics teaching, the focus must be on building meaning (Davis, 1992), and not just on imitation or recall — both for teachers as well as for students. The teacher development approach described here is designed to help teachers build "understanding" of content, pedagogy, and students' thinking, which in turn can help students build understanding. The participating teachers came from urban as well as suburban school communities, and they were able to revise their ways of thinking about the teaching and learning of mathematics. They changed their perceptions regarding the most important things to observe when students are engaged in problem activities, their views on how to help students reflect on and assess their own work, and their ideas about what curricula materials would be effective in helping students build mathematical ideas. Of prime importance, however, is the fact that the nature of their students' thinking changed as well. As the examples reveal, the students' work was initially superficial, although most of the teachers did not realize that. It was only after sustained and in-depth experiences that the teachers began to notice the depth in their students' thinking. The more they noticed, the more thinking their students actually did.

References

Ball, D. L. (2001). Teaching with respect to mathematics and students. In T. Wood, B. S. Nelson, and J. Warfield (Eds.), *Beyond classical pedagogy* (pp. 11-26). Mahwah, NJ: Lawrence Erlbaum.

Cobb, P., Wood T., Yackel, E., & McNeal, E. (1993). Mathematics as procedural instructions and mathematics as meaningful activity: The reality of teaching for understanding. In R. B. Davis and C. A. Maher (Eds.), *Schools, mathematics, and the world of reality* (pp. 119-133). Boston: Allyn and Bacon.

Cuban, L. (1993). *How teachers taught: constancy and change in American classrooms, 1890-1980.* (2nd ed.). New York: Teachers College Press.

Davis, R. B. (1992). Understanding "understanding." *Journal of Mathematical Behavior, 11,* 225-241.

Davis, R. B., & Maher, C. A. (1997). How students think: The role of representations. In L. English (Ed.), *Mathematical reasoning: Analogies, metaphors, and images* (pp. 93-115). Mahwah, NJ: Lawrence Erlbaum.

Fennema, E., Sowder, J., & Carpenter, T. A. (1999). Creating classrooms that promote understanding. In E. Fennema and T. A. Romberg (Eds.), *Mathematics classrooms that promote understanding* (pp. 185-199). Mahwah, NJ: Lawrence Erlbaum.

Firestone, W. A., Schorr, R. Y., & Monfils, L. A. (2004). *The ambiguity of teaching to the test.* Mahwah, NJ: Lawrence Erlbaum.

Franke, M. L., & Kazemi, E. (2001). Teaching as learning within a community of practice. In T. Wood, B. S. Nelson, and J. Warfield (Eds.), *Beyond classical pedagogy* (pp. 47-74). Mahwah, NJ: Lawrence Erlbaum.

Goldsmith, L. T., & Schifter, D. (1993). Characteristics of a model for the development of mathematics teaching. *Reports and papers in progress.* Newton, MA: Center for Learning, Teaching & Technology, Education Development Center.

Lesh, R., Amit, M., & Schorr, R. Y. (1997). Using real life problems to prompt students to construct conceptual models for statistical reasoning. In I. Gal and J. B. Garfield (Eds.), *The assessment challenge in statistics education* (pp. 64-84). Amsterdam, NL: IOS Press.

Lesh, R., & Doerr, H. (2003). In what ways does a models and modeling perspective move beyond constructivism. In R. Lesh and H. Doerr (Eds.), *Beyond constructivism: a models and modeling perspective on teaching, learning, and problem solving in mathematics education* (pp. 519-556). Mahwah, NJ: Lawrence Erlbaum.

Lesh, R., Hoover, M., & Kelly, A. E. (1992). Equity, assessment, and thinking mathematically: principles for the design of model-eliciting activities. In I. Wirszup and R. Streit (Eds.), *Proceedings of the UCSMP International Conference on Mathematics Education: Developments in School Mathematics Education Around the World: Volume 3*, (pp. 104-129). Reston, VA: National Council of Teachers of Mathematics.

Ma, L. (1999). *Knowing and teaching elementary school mathematics: Teachers' understanding of fundamental mathematics in China and the United States*. Mahwah, NJ: Lawrence Erlbaum.

National Council of Teachers of Mathematics. (2000). *Principles and standards for school mathematics*. Reston, VA: Author.

Schorr, R. Y. (2000). Impact at the student level. *Journal of Mathematical Behavior, 19*, 209-231.

Schorr, R. Y., & Firestone, W. A. (2004). Conclusions. In W. A. Firestone, R. Y. Schorr, and L. F. Monfils (Eds.), *The ambiguity of teaching to the Test* (pp. 159-168). Mahwah, NJ: Lawrence Erlbaum.

Schorr, R. Y., Firestone, W. A., & Monfils, L. (2003). State testing and mathematics teaching in New Jersey: the effects of a test without other supports. *Journal for Research in Mathematics Education, 34*(5), 373-405.

Schorr, R. Y., & Koellner-Clark, K. (2003). Using a modeling approach to consider the ways in which teachers consider new ways to teach mathematics. *Journal of Mathematical Thinking and Learning, 5*(2), 191-210.

Schorr, R. Y., & Lesh, R. (2003). A models and modeling perspective on classroom-based teacher development. In R. Lesh and H. Doerr (Eds.), *Beyond constructivism: a models and modeling perspective on teaching, learning, and problem solving in mathematics education*, (pp. 141-157). Mahwah, NJ: Lawrence Erlbaum.

Shafer, M., & Romberg, T. A. (1999). Assessment in classrooms that promote understanding. In E. Fennema and T. A. Romberg (Eds.), *Mathematics classrooms that promote understanding* (pp. 159-184). Mahwah, NJ: Lawrence Erlbaum.

Simon, M. A., & Tzur, R. (1999). Exploring the teacher's perspective from the researchers' perspectives: generating accounts of mathematics teachers' practice. *Journal for Research in Mathematics Education, 30*(3), 252-264.

Simon, M. A., Tzur, R. Heinz, K., Kinzel, M., & Smith, M. S. (2000). Characterizing a perspective underlying the practice of mathematics teachers in transition. *Journal for Research in Mathematics Education, 31*(5), 579-601.

Sowder, J., & Philipp, R. (1999). Promoting learning in middle-grades mathematics. In E. Fennema and T. A. Romberg (Eds.), *Mathematics classrooms that promote understanding* (pp. 89-108). Mahwah, NJ: Lawrence Erlbaum.

Spillane, J. P., & Zeuli, J. S. (1999). Reform and teaching: Exploring patterns of practice in the context of national and state mathematics reforms. *Educational Evaluation and Policy Analysis, 21*(1), 1-27.

Roberta Y. Schorr, Associate Professor of Mathematics Education at Rutgers, the State University of New Jersey Campus at Newark, is the Principal Investigator or Co-Principal Investigator on several grants designed to study and improve the teaching and learning of mathematics. Her research, teaching, and service have three interrelated goals, all focusing on the ultimate objective – the teaching and learning of powerful mathematics, with understanding, for all students. Dr. Schorr has written articles on these topics, and is the co-editor (with W. Firestone and L. Monfils) of a book entitled, *The Ambiguity of Teaching to the Test*. [Schorr@rci.rutgers.edu]

Printed in the United States
By Bookmasters